Anonymous

Mines of the Pacific Coast

Anonymous

Mines of the Pacific Coast

ISBN/EAN: 9783744793896

Printed in Europe, USA, Canada, Australia, Japan

Cover: Foto ©berggeist007 / pixelio.de

More available books at **www.hansebooks.com**

CHRONICLE, TUESDAY, MAY 24, 1892.

MINES OF THE PACIFIC COAST

The recent action taken by Congress, looking to a revival of the gold mining industry in this State, has aroused public interest in the subject of the supply of the precious metals, and makes this an opportune time for summarizing the facts in regard thereto, particularly with reference to the great mineral region of the West. The wonderful discoveries of gold and silver made in this region during the last half century have surpassed anything recorded theretofore in the world's history. That these discoveries revolutionized financial methods need not be more than pointed out. That they added more to the world's wealth than any other fifty years since the dawn of creation will not be denied. That the deposits, instead of approaching exhaustion, have, in fact, only been exploited to an inconsiderable degree, is firmly believed by every miner who has made the subject a study. The surface has been skimmed over, a small way. Along the seashore in Humboldt and Del Norte counties, formerly the chief sites of this class of mining, the residents of that section of the State gather from these sands by hand sluicing a little gold every year. Their earnings are small and their labor intermittent, being prosecuted only when they have water for the washing, which in most localities is the case during only a small portion of the year.

Besides these "gold bluffs" and "beaches" we have in California a variety of other auriferous deposits, some of which, like the gold bluffs, are peculiar to the State; nor do more than a few of the others meet elsewhere with such large development as here. The principal of these deposits, designating them by the local names, consist of the following, viz.:

The dry diggings, so called, are simply such surface placers as, being without a sufficient natural supply of water for washing, cannot be supplied by artificial means. There are many localities of this character in California. In cases of this kind, if the auriferous earth is not rich enough to bear transportation to water, the gold is separated from it by "dry washing," a process formerly conducted by means of the Mexican batea, still employed in some places. By the Spanish-speaking races the batea continues to be exclusively used in the dry diggings, and these people are very skillful in handling it. Latterly dry-washing machines of various kinds have been invented, some of which are efficient, as much so, in fact, as can reasonably be looked for consider-

"urce of supply will in time have to be systematically exploited in order to keep pace therewith.

It is certain that the gold production of the world is steadily decreasing, while there is a constantly increasing demand for purposes of ornamentation. Not more than $100,000,000 worth is now mined annually, and that is not enough to meet the demand, as shown by the constantly increasing value of the metal, as evidenced by its increased purchasing power. The available mines of the world are being rapidly exhausted, while the unexplored portion of the world's surface grows less in extent each year and the possibility of finding new mines becomes less promising. In ancient times gold was obtained abundantly from the rivers of Asia. The sands of Pactolus, the yellow metal of Ophir, the fable of King Midas, all illustrate the Eastern origin of gold. Alexander the Great brought nearly $500,000,000 of gold from Persia. Gold also came from Arabia and from the middle of Africa by way of the Nile. But all of these sources of supply were long since exhausted. Brazil, which a century ago was a rich gold producing country, has ceased to be so, and the famous Gold Coast of Africa has lost its productiveness. Since the commencement of the sixteenth century Africa has produced $500,000,000 worth of the precious metal, but little is now obtained there. Australia has yielded $1,300,000,000 worth of gold, but the production has greatly decreased. Not less than seven billion dollars' worth of gold has been dug in the world since the discovery of America, but, nevertheless, the world's supply is becoming so scarce that the yellow metal will undoubtedly soon be hoarded to such an extent that before the expiration of many centuries it will have attained a value several times greater than at present.

The importance of exploiting all the available sources of supply in California and elsewhere on the Pacific coast is emphasized by these undisputed facts. The belief is widespread that our gold mines have been exhausted, and little is to be expected from them in the future. This those who have examined the matter know to be an erroneous conclusion. They know that there are undeveloped sources which will in time yield as great an amount as has yet been produced, if, indeed, they do not largely surpass it.

From the last report of the Director of the Mint, covering the year 1891, the following facts are taken: "The product of gold from the mines of the United States aggregated 1,604,840 fine ounces, of the value of $33,175,000. This is an increase of $330,000 over the product of the previous calendar year. The increased product is due largely to improved processed of treatment and to the increased amount of gold extracted from lead and copper ores.

The product of silver from our own mines was 58,330,000 fine ounces, of the commercial value of $57,630,040 or of the coinage value in silver dollars of $75,416,565. This is an increase of 3,830,000 ounces over the previous year. The increased silver product was due principally to new finds in Colorado and Idaho, and to the cheapening of the process of smelting lead and copper ores bearing silver.

The Director of the Mint has made special effort to distribute for the first time the silver product of the United States as to the sources of production. He estimates that of the total product for the last calendar year 28,497,000 fine ounces were produced from quartz and milling ores, 23,707,000 from lead ores and 6,126,000 from copper ores. Total silver output, 58,330,000 fine ounces.

The total product of Government and private refineries in the United States, including foreign material smelted and refined, was: Gold, 2,169,863 fine ounces; silver, 69,336,415 fine ounces.

The total value of the gold deposited at the mints during the year was $70,915,632, of which $24,853,180 was foreign coin and bullion. The deposits and purchases of silver aggregated 73,088,626 standard ounces, of the coining value of $85,048,584. The amount of silver purchased by the Government during the year was 54,393,912 fine ounces, costing $53,796,833. The average cost of the silver purchased during the year was $0.989 per fine ounce. The average cost of the total amount purchased under the act of July 14th, 1890, has been $1.02 per fine ounce."

The price of silver at the commencement of the calendar year 1891 was $1.058 per fine ounce, and at the close, December 31st, was $0.955 per fine ounce. The average price for the calendar year was $0.99 per fine ounce.

At the date of the passage of the act of July 14, 1890, the price of silver was $1 07¼ per fine ounce; at the date the law went into effect it had advanced to $1 13. The highest point touched was on August 19, 1890—$1 21 per fine ounce. The lowest point reached was on March 28, 1892—$0.85½ per fine ounce.

According to the reports of the transportation companies the bullion product of the States and Territories west of the Missouri river for 1891 was as follows:

Alaska....$	850,000	Montana..$	28,011,000
Arizona....	5,576,157	Nevada....	8,745,611
California.	12,215,233	New Mex..	4,237,740
Colorado...	28,204,037	Oregon....	1,084,000
Dakota.....	3,422,871	Utah......	13,408,493
Idaho......	11,595,000	Wash'gton	320,000

These figures, however, do not represent the entire production to a large extent. Much bullion is carried otherwise than by the express companies, while vast quantities of ore are shipped for treatment to outside points, the product of which does not appear in such reports as that quoted. As will appear further on, these figures require considerable revision in order to arrive at the actual production of the various States and Territories.

In the succeeding columns the various sources of production will be pointed out, the more notable districts and mines will be described in detail and an effort will be made toward removing the widespread misapprehension that exists upon this subject, that the mines of the Pacific coast are "played out," and that further effort in this direction is useless.

It will be shown that from Alaska to the Mexican line and from the Pacific to the eastern slopes of the Rocky mountains are mineral belts of vast extent which have as yet scarcely more than begun to give up their wealth. It will be shown that besides the vast deposits of gold and silver bearing rock, there is a store of other minerals of a diversity not found in any other part of the world. It will be shown that there are opportunities for investment and for the exercise of energy and ability that equal, even excel, those that have made fabulous fortunes in the past.

All this and more, too, will be given in as plain and straightforward a manner as possible, for the purpose of educating the people to a correct knowledge of the great wealth that may be theirs for the taking.

CALIFORNIA.

HER VAST DEPOSITS OF THE PRECIOUS METALS.

Where Gold Was First Discovered—Different Methods of Mining—Valuable Silver Mines — Copper, Coal, Quicksilver, Etc. — Asphaltum and Petroleum — The Cajalco or San Jacinto Tin Mines.

In the variety and extent of her mineral wealth California has scarcely a rival and certainly no superior. The popular conception in regard to this matter is that the principal if not the sole resource of this character possessed by this State is the deposits of gold, which the majority of people outside its boundaries undoubtedly believe are nearly or quite exhausted. In both ideas they are mistaken. The gold mines of California will yet yield, it is the opinion of those who have made the subject a study, fully as much if not many times more than the amount of treasure that has already been delved from them. But in addition this State possesses latent mineral wealth of the most surpassing and extensive variety. To prove this it is only necessary to mention the fact that the range of deposits includes silver, quicksilver, copper, tin, iron, lead, coal, antimony, asbestos, sulphur, borax, soda, petroleum, asphaltum, and a host of other substances of more or less value, the exploitation of which is certain to add millions to the wealth which they have already created.

From whatever standpoint the mineral wealth of California be considered, the subject is one of interest and always will remain so. True, the romance of the early gold-mining days is past, never to return, and the search for the golden treasure has become a prosaic industry similar to other productive enterprises. Nevertheless there is always an interest about the contest for the contents of nature's treasure box that makes the subject one of perennial freshness.

GOLD MINES.

Where the Yellow Metal Was First Discovered—Fears of a Plethora.

The history of the discovery of gold in 1848 in California has been so frequently told and the facts are so well established that there is nothing of interest to be added to the well known and familiar account of the Coloma Mill, the finding of the particles of gold in the tail-race, and the subsequent operations of General Sutter, Marshall and the others who were present or were at once apprised of the discovery. The story has been told a thousand times and is familiar the world over.

It is not so well known, however, that, while Marshall's discovery was unquestionably the one that produced the most wonderful migration and subsequent development of an unknown region that the world has ever seen, he is by no means entitled to the honor of having been the first person to find the precious metal in California.

Nothing can be more assured than the fact that from almost the first exploration of the Pacific coast by the hardy navigators of the sixteenth century, the idea in some way gained a foothold that gold existed here in abundance. Sir Francis Drake, who visited this region in 1579, asserts it, and so do other writers who have other sources of information. The Spanish conquerors of Mexico were persuaded of the existence of rich gold deposits in a country far to the northwest, corresponding exactly with the location of our State, but were unable to verify their belief, though sending out frequent expeditions to do so.

That the founders of the missions knew of the existence of gold here there is good ground for believing, as well as for believing that they profited by that knowledge.

In 1775 gold was discovered near the Colorado river in the vicinity of Yuma by Mexicans, and half a century later deposits were found near San Ysidro, in San Diego county. In 1833 placers which are still being successfully worked were found in the mountains to the northwest of Los Angeles, and from them were taken considerable quantities of the precious metal. Some of the product of these mines found its way to the Atlantic seaboard long before Marshall was ever heard of, and the knowledge of the existence of gold on the Pacific coast was quite general even then. This fact was known to the Mexican authorities as early as 1844, as shown by documents found in the archives of that Government. In one communication, dated September 1, 184 it was said that fully 2000 ounces of gold dust taken from the placers of the Santa Clara were in circulation at one time in Los Angeles, and in the same letter the existence of silver mines is also mentioned, though their exact location is not given.

In March, 1846, nearly two years before the discovery at Coloma, Thomas Larkin, Consul at Monterey, wrote to his superiors that he had no doubt that mines of gold, quicksilver, copper, etc., would be found all over California. Five years before that J. D. Dana, who accompanied the Wilkes expedition and made an overland trip from Oregon to San Francisco, reported that he found indications of the existence of gold in Southern Oregon and in the Sacramento valley. Many other facts might be cited, all tending to establish the certainty that the discovery of Marshall was no discovery at all in the real sense of the word, though by a fortuitous combination of circumstances his lucky (or rather unlucky for himself) find set the world in a blaze of excitement.

Not only were the people of every civilized land carried away by the tales of great fortunes to be made in a day, but the financial and monetary world was appalled and shaken to the base by California's extraordinary output of the precious metal. Europe became alarmed. A plethora of the noble metal was feared, and for a time the idea was strongly entertained of demonetizing gold.

Primitive Mining Methods.

The yield of gold was something extraordinary. At first the general gains of the miners, though great, were small compared to what shortly afterward were collected. By comparing different accounts and endeavoring to form from them something like a fair average, it is found that from $10 to $15 worth of gold dust was at first about the usual proceeds of an ordinary day's work. But while that might have been the average, well authenticated accounts describe many persons as averaging from $100 to $200 a day for a long period, and numerous others are said to have earned as high as $500 to $800 a day. If, indeed, a man with a pick and pan did not make a fortune rapidly he moved off to some place which he supposed might be richer. When the miners knew a little better about the business and the mode of turning their labor to account the returns were corresponding increased. At what were called the "dry diggings," particularly, the yield of gold was simply enormous. One nugget of pure metal was found of thirteen pounds weight. The Johnson instrument at first made use of was a butcher's knife. Afterward the pick and shovel were used. The auriferous earth, dug out of ravines and holes in the sides of the mountains was packed on horses for one, two or three miles to the nearest water to be washed. An average price of this washing dirt was $400 a cartload. In one instance five loads sold for $752, which, after washing, yielded $16,000. Cases occurred where men carried the earth in sacks on their backs to the watering places and collected $800 as the proceeds of their labor. Individuals made their $5000, $10,000 and $15,000 in the space of only a few weeks. One man dug out $12,000 in about six days. Three others obtained $8000 in a single day. But these, of course, were extreme cases. Still, it is undoubtedly true that a large proportion of the miners earned such sums as they had never seen in their lives, and which six months before would have appeared like the wildest fable.

The washing was effected by putting the earth in a pan or bowl, mixing water with it and violently shaking the contents. A peculiar shake of the wrist, best understood and learned by practice, threw the were built and magnificent roads laid. By the use of ingenious contrivances water was given a pressure sometimes as high as 500 feet and a velocity of 160 feet per second. With this, equal to the force of a small Niagara, the base of the hills was washed away and the summit toppled over like a building undermined. Great rocks of hundreds of pounds weight were tossed about like straws in the current. Whole mountains were moved in this way and the very topography of the country changed.

It is interesting to note here that while we take the credit to ourselves of having invented hydraulic sluicing, our mighty nozzle work was but an exaggeration of the process used by the Romans in Spain. Thus Pliny writes: "Another labor, too, quite equal to this, and one which entails even greater expense; is that of bringing rivers from the more elevated mountain heights, a distance in many instances of 100 miles, perhaps, for the purpose of washing these debris. Then, too, valleys and crevasses have been united by the aid of aqueducts, and in another place impassable rocks have to be hewn away and forced to make room for hollow troughs of wood. The earth carried onward in the stream arrives at the sea at last, and thus is the shattered mountain washed away, causes which have greatly tended to extend the shores of Spain by these encroachments upon the deep."

The history of hydraulic mining in California, when it comes to be written up, will be full of wonders. So long as the rich placers lasted there was little inducement to seek for their origin; but as they declined the more enterprising of the miners commenced tracing these alluvial deposits to their sources. The researches thus undertaken led to some remarkable and astonishing discoveries. In many instances the gravel, being worked in open river beds, was found to burrow abruptly into the sides of high mountains, and then it was realized that the stream which had accumulated the treasure belonged to a past geological period and that its bed had been filled ages ago by a stream of very different character—a solid instead of a liquid stream; in other words, a lava flow. Numerous instances have occurred where such an extinct river bed has received successive lava flows, one superimposed upon another, with auriferous gravel between, showing that the river resumed, as nearly as might be, its original channel after each invasion of molten rock.

The yield of gold from these ancient

---

and to all of us, of the do Library Building, which, tho means the least, among the thought and generous impul is at this moment and in th manifestation of that sentim to express our thanks, and fo incoming ages, we foreshado tions of students of the fu advantages, the seeds of whi us.

I said a moment ago tha all the citizens of the State ing so, I spoke advisedly, you all feel, that this is the founded by the people, for

For a few of the moment your attention, I shall ask

streams, locally known as "dead rivers"—a most apt expression—has been immense, for they must have been mighty floods, draining huge areas, and during their long and active lives they were ceaselessly helping to accumulate the scattered riches contained in the surrounding rocks, these riches being liberated by the action of frost and thaw and rain and snow and sun, whose combined effect disintegrated the quartz veins that carried the gold. Thus Nature, working in her own slow and secret way, collected into comparatively narrow limits, ready for the use of man, the gold which had been disseminated through millions of tons of rock, probably in such small proportions as not to repay the cost of extraction by human methods. More than that, the precious metal actually underwent a certain degree of refining at the same time, the accompanying base metals having been dissolved out and washed away.

Hydraulic Mining.

Hydraulic mining added largely to our annual output until in 1876 litigation commenced between the farmer and the miner. A bitter fight in our courts ensued, which resulted in favor of the agriculturists. This was followed by the appointment of a commission of engineers to investigate the subject from an engineering standpoint and report.

For years there has been a practical interdiction of hydraulic mining except in a few remote localities, and many millions of dollars have been lost to the people of this State. Finally, however, owing to the discoveries of the engineers in charge of the investigation, it has become apparent that it is possible to bring about a resumption of the working of these valuable deposits, and from present appearances it will not be long before the foothills of the Sierra will again be contributing their golden wealth to the industries of the State.

The importance of hydraulic mining may be seen from the fact that it is estimated that of the entire gold product of this State at least nine-tenths was yielded by the auriferous gravels. The total yield thus obtained would be represented by a cube fourteen feet square. These auriferous gravels occur in the channels of ancient rivers, and there are 400 miles of these, which at a low estimate will yield $2,000,000 to $3,000,000 to the mile.

According to the reports of the engineers detailed by the Government to examine into the question of mining debris, there were some 857,000,000 cubic yards of material excavated during the prevalence of hydraulic operations, of which 280,000,000 remained in the beds of the three principal rivers affected—the Yuba, Bear and American. After a careful examination of the damage done by this debris the engineers reported the following as the injury done along the three streams where the greatest amount of loss was caused:

Name.	Destroyed, acres.	Loss	Injured, acres.	Amount.
Feather river..	17,628	$1,097,038	6,940	$196.750
Yuba river.....	11,845	1,079,577	3,500	144,500
Bear river.....	9,741	694,970	3,515	81,200
Total........	39,214	$2,871,585	13,955	$422,450

It is conceded and was demonstrated to the board of engineers that certain lands are capable of being improved by the addition of small quantities of slickens. It is also stated that some lands are benefited by the rising of the adjacent water, which makes them moist and cultivable. The extent of these favorable features was not possible of determination.

	Acres.
Area destroyed as above.	39,214
Area injured as above.	13,955
Total area	52,169
Value of land destroyed	$2,871,585
Value of land injured	422,450
Total loss	$3,304,035

There can be no doubt that the miners have contributed to the filling of the mining rivers ever since mining commenced in California, and that the people whose lands have been covered by debris have a right to complain, and had they when the evil first commenced taken proper measures the money value of their injury could have been compensated. It was not until the flood of 1861-62 swept down into the lower stream the thirteen or fourteen years' accumulation in the mountains of mining debris that the evil began to be very injurious; that every creek, gulch, stream, canyon or bar was up to that time swarming with miners is well known. Perhaps no better evidence of the fact can be shown than the yield of gold during the intervals of time between 1848 and including 1861:

From 1848 to and including 1849 it was only...............	$10,306,661
From 1849 to and including 1854 it was...................	335,553,456
From 1854 to and including 1859 it was...................	249,060,717
From 1859 to and including 1861 it was only............	35,080,158
Total from commencement of 1848 to close of 1861	$680,990,992

The largest yields were in 1851, $75,938,-232, and in 1852, $61,294,700. The yield for 1866 was only $12,579,856, being the

THE MAN WHO DISCOVERED GOLD.

...est yield ever known. The total yield, so far as known, has been $1,144, 64,521, but it is believed that the actual yield has been in excess of this sum, certainly up to $1,200,000,000.

During the first thirteen years, or up to the time when the flood of 1861-62 filled the mining rivers, more than half of the total product (or $680,000,000) was extracted, while during the twenty-five succeeding years some $466,000,000 only was extracted. Hydraulic mining did not commence on a large scale until about 1867, although it was some years after that date before it assumed the proportions of 1880. Prior to 1867 it was carried on upon a very limited scale.

The myriads of miners at work on the slope of the Sierra deposited their tailings —all of light character—into the streams adjacent to where they worked, for water was scarce and expensive, and as every miner so disposed of his tailings as not to deposit them upon the claim below him, these vast quantities accumulated until the flood of 1861-62 swept them all into the rivers and the evils now complained of them became of serious nature. After this time the miners on the Yuba contributed $80,000, unasked, to aid in building levees along the south side of that river above Marysville; so that all the evils now complained of are not chargeable to the hydraulic miners. It is, however, upon the heads of the present miners that the sins of nearly forty years now fall, to their ruin and to their loss to an extent of over $100,000,000.

The famous decision of Judge Sawyer, under which hydraulic mining was suspended, contained the following clause, the case being that of Woodruff vs. the North Bloomfield Mining Company et al.:

"On consideration whereof it is by the court ordered, adjudged and decreed as follows, to wit: That the defendants herein and their and each and all of their servants, agents and employes are perpetually enjoined and restrained from discharging or dumping into the Yuba or into any of the forks or branches or into any stream tributary to said river or any of its forks, ravines or branches, and especially into Deer creek, Sucker Flat ravine, Humbug creek, Scotchman's creek, any of the tailings, bowlders, cobblestones, gravel, sand, clay, debris or refuse matter from any of the tracts of mineral land or mines described in the complaint; and also from causing or suffering to flow into said rivers, creeks or tributary streams aforesaid therefrom any of the tailings, bowlders, cobble-stones, gravel, sand, clay, or refuse matter resulting or arising from mining thereon; and also from allowing others to use the water supply of said several mines or mining claims or any part thereof for the purpose of washing into said rivers and streams any earth, rock, bowlders, clay, sand or solid material contained in any placer or gravel ground or mine."

As a result of this inhibition a product of $10,000,000 annually was cut off, a large share of which had found its way directly into the channels of trade. At the same time property in which had been invested fully $100,000,000 was made useless, and has remained so to this time.

In reply to the question, "Can hydraulic mining be resumed without injury to the navigable streams?" the Board of Engineers reported:

"It is not apparent to the board that any expression of opinion or recommendation will have any effect in rehabilitating the industry in the present legal status of the question. Without some modification, then, of existing conditions hydraulic mining must cease. It cannot be carried on without violating the decrees of the courts.

"If, however, by a reversal of the opinions of the courts or by other means hydraulic mining be permitted in whole or in part, or if without such reversal an expression of opinion is required as to the feasibility of impounding mining debris, the board will state that the investigations and examinations made indicate that in isolated cases it is possible to impound debris without injury; also, that locations exist in the canyons of the different mining streams in the Sierra district where permanent stone dams, properly constructed, will retain large quantities of material of the character formerly mined out and which caused the destruction of the farming lands and injured the navigation of the rivers.

"These dams, however, will not be effective in impounding all the material delivered into the canyons from the mines. Being in the streams and in the pathway of the freshets, portions of the heavier material will be carried over the crests of the dams to eventually find lodgment in the river below. The finer sands and clays cannot be effectually impounted by such barriers, but will be carried off in suspension. With the improved condition which it is desired to give to the navigable rivers, it is probable that the greater part of this finer material can be carried off without being productive of harm."

A detailed statement is made by the chief of the corps of engineers, in which the location of the impounding dams desirable to be constructed is pointed out, together with their cost and the amount of debris capable of being restrained thereby. It is estimated that by an average annual expenditure of $300,000 for eight years fully $10,000,000 each year may be taken from the mines. At a moderate calculation there remain in the known auriferous deposits over 2,100,000,000 cubic yards of gravel, and this at a low rate will yield over $552,000,000. These facts, which are well established, show the vast importance to California of the reopening of these mines.

John H. Hammond, a prominent mining engineer and expert, estimates that there are available for hydraulic working deposits that contain fully $800,000,000, while there are in the ancient lava-capped channels fully $500,000,000, or a total of $1,300,000,000.

In accordance with the recommendations of the board of engineers a bill is now before Congress providing for the commencement of operations upon impounding claims which shall enable the mines to be again worked.

Although washing by the hydraulic method has been enjoined in the central mining counties, formerly the field of its largest operations, it is still carried on in the northwestern part of the State, chiefly in Del Norte, Trinity, Humboldt and Siskiyou counties. In that region there exists no objection to its being prosecuted, while the conditions for doing so are exceptionally good. All included, there are in this group of counties not less than fifty hydraulic claims being operated at the present time, the most of them, however, only for a portion of the year, and in a small way. They nearly all make liberal returns for the labor employed and the amount of money expended in fitting them up; the latter is not generally large, as lumber is cheap, and no costly bedrock tunnels are ever required, while comparatively short ditches suffice to introduce water on the ground to be washed. The auriferous gravel banks throughout this region are generally large, the material being at the same time of good grade, and free from pipe clay and other barren matter. There is everywhere fall enough to prevent any troublesome accumulation of tailings below the washing pits, and there being no farming lands along the outletting streams liable to be injured by the debris from the mines, there is apparently no reason why hy-

...here for an indefinite period, and with large profits.

It was in this section of country that the style of hydraulic mining known as "booming" was first introduced, and has since been most largely used. It is practiced only along the gulches. These affording but little water, it became necessary that the limited supply be reservoired and properly distributed in order to make it effective in this method of gravel washing. The object is attained by retaining the water in dams and then releasing it suddenly, with a rush or boom. Near the bottom of the dam built for this purpose is left an aperture so large that when opened the water escapes rapidly. Placed on the top of the structure is a small race, through which the water flows when the dam is full, and is discharged into a large wooden box suspended from the end of the sweep, turning on a pivot, and the upper end of which extends to and over the top of the dam. Attached to this end of the sweep is a strip of heavy canvas, which, dropping in a fold over the aperture below, keeps it tightly closed when the dam is full.

When this stage has been reached the water flowing through the race into the wooden box mentioned soon fills it, causing this end of the sweep to sink and the other end to rise, carrying with it the strip of canvas and uncovering the large aperture below, allowing the water to rush out. Meantime, the wooden box having emptied itself through numerous small holes made for the purpose, this end of the sweep, relieved of its weight, rises and the other end drops. The canvas falls over the outletting aperture, closing it as before. Then the dam fills again to the brim and the operation as above is repeated. This plan for handling water is wholly automatic. It takes care of itself and goes on day and night without any attention on the part of the miner, doing its work as long as the water lasts. This is one of those ingenious contrivances for which the California miners have ever been noted. Since its introduction in the northwestern part of the State it has been brought into use in many other places, some of which have presumed to claim its paternity, a distinction that unquestionably belongs to this State, where this device was originally known as the "self-shooter." While to us belongs the credit of this invention, to others belongs the credit of having substituted for the above name the more appropriate one by

Drift Mining.

During the period of cessation of hydraulic mining attention was largely directed to other methods. These are drift mining, quartz mining and river-bed mining. Drift mining, which seems to have gained a remarkable impetus since the suspension of hydraulic mining, is conducted as follows: The prospector having come to the conclusion that there is a bed or deposit of gold-bearing dirt, quartz or gravel within the recesses of a certain hill then seeks the easiest way to get at it. If a vertical shaft from the top of the mountain be considered the shortest direct road to the treasure then such a shaft is sunk. If the pay dirt is thought to be best reached by a horizontal tunnel through the side of the mountain then such a tunnel or drift is run, with such ramifications or drifts as occasion may call for. The ore taken out is treated according to its character. It is a comparatively cheap form of mining, and so far it has been found just as remunerative to honey-comb a mountain as to wash it away. Drift mining, now comparatively in its infancy, is bound to assume considerable prominence. It is of most importance now in Placer, Nevada and Sierra counties. It really is a revival, having been pursued to a considerable extent early in the history of the State, and then abandoned for the hydraulic style. During the past ten or a dozen years, however, it has been resumed with very satisfactory results. Already it has done a good deal to replace the millions added to the annual production of the State under the old system, and drift mining will in future years add scores of millions to the wealth of the world. Drift mining, indeed, is now regarded as about the most safe and certain branch of the business extant. Through the employment of machine drills and more powerful explosives both the cost and length of time required for opening this class of deposits have been greatly reduced. The engineering difficulties that formerly attended this work have also been much diminished, a better acquaintance with the position of the old channels making it possible now to drive the exploiting tunnels almost always on the right level.

This class of deposits has come to be sought after, and where open to location are speedily taken up. A vast amount of exploratory work has been projected and much commenced, a large number of shafts and tunnels being in operation. Work long suspended on partially completed structures has been resumed, while operations on the productive mines are being pushed with energy. Some of these drift mines already employ from 100 to 200 men, their gross yearly output varying from $150,000 to $300,000. Most of the claims, however, are operated with a much smaller working force, the number of men employed ranging from ten to fifty, the production being correspondingly small. The deposits sought by drifting rest for the most part in the "dead river" channels before referred to.

The Forest Hill divide in Placer county, the Magalia district in Butte county and the vicinity of Forest City in Sierra county continue the most active and largely productive drift localities. A good deal is also being done in this line of mining along the Liberty Hill ridge, about Nevada City, near Gibsonville and at other points in Western Sierra. Some very heavy operations of this kind have recently been set on foot in the latter locality.

Since the suppression of hydraulic washing in the central mining counties of the State a number of claims before operated by that process have been worked by drifting, and in most cases with satisfactory results. In a few instances, however, these attempts proved so disappointing that they have been abandoned.

No very heavy drift operations are carried on in the extreme northern part of the State, nor in any of the counties south of Tuolumne, for the reason that in neither do the rich pliocene deposits occur. In the tier of counties north of Plumas and Butte drift mining is prosecuted at a great many different points, but mostly in a limited way and along the banks of present streams or in the buried river channels of a comparatively recent date. In the regions mentioned the claims worked in this manner are so limited in extent and their product so unimportant that they scarcely require to be individually mentioned. The working force employed is invariably small, rarely ever exceeding ten or fifteen hands, and usually about half that number.

Coming south into Butte and Plumas counties we enter a very extensive and productive field of drift mining. The "dead rivers" here appear in great strength. The Spring Valley Company at Cherokee, Butte county, having worked their ground for many years with notable success by the hydraulic process, concluded to abandon that plan and adopt the drift method. The increasing depth of the superincumbent volcanic matter is the cause of this change. Surveys for the new tunnel have been completed and an estimate of the expenditure required for effecting the contemplated change also made. The expense is much less than was at first anticipated. This mine has for the past twenty-five years employed an average of 250 men; its output of gold amounted, meantime, to fully $10,000,000. To make this production only about 4800 linear feet of the channel included within the company's ground have been exhausted. Over 4000 feet still remain, but this being of extra large dimensions—800 feet wide and nine feet thick—it is believed the remaining section, worked by the more economic drift method, will yield a total of $15,000,000. As the company own over 2000 inches of water and only about 300 inches are required for drifting, they will have a large surplus to be sold for irrigation purposes. The principal drift claims worked in this county are the Magalia, Lucretia, Bay State, Oro Fino, Indian Springs and Eureka, several of lesser importance having been operated in the vicinity of Little and Big Butte creeks.

The value of the gravel extracted in Butte ranges from $1 to $5 per carload; mean value, about $2. In thickness the stratum removed ranges from two and a half to four feet and in width from twenty to seventy-five feet, that in the Cherokee ground having the exceptional average thickness of nine feet and width of 600 feet. The number of men employed in these mines runs from five to fifty, the average being not above ten. The ground is opened by tunnels, but as some of these have been run on too high a level to effect complete drainage the water has to be lifted to the tunnel level by pumping. Nearly all the old channels in Butte are lava-capped, and they have to be worked by drifting, hydraulic washing being practicable in only a few localities.

Although Plumas is not largely a drift county, it contains several good claims of this class, the Sunny South and the Glazier being the most prominent. A portion of the North America has an entrance in Sierra and extends over the line into Plumas.

Coming into Sierra county, we arrive at the heart of the northern drift mines, with Forest City for its center. At this place two large companies, the Bald Mountain Extension and the South Fork, are actively operating. The North America Company, at the head of Slate creek, have been drifting to open up new ground, of which they have large reserves supposed to be rich. The channels worked out have yielded generously for the past twenty years.

Monte Cristo and Port Wine, famous old drift camps, are likely, through the investment of much capital in their vicinity—the most of it English—to soon regain their former importance? Some of these new enterprises are already producing handsomely, and they promise to largely increase the amount in the future. Additional ground has been bonded by these and other foreign companies, and it may be expected that the drift industry will at no distant day be brought into a flourishing condition along with the entire Monte Cristo gravel range.

There are several prosperous drift camps in Nevada and Placer. In the vicinity of Red Dog and You Bet this style of gold mining gave profitable employment to hundreds of men many years ago. The business afterward fell into decadence, and it is now undergoing marked restoration.

In other parts of the county a number of drift claims have in like manner been resuscitated and are now successfully operated for the first time in several years. The Manzanita, near Nevada City, a large and steady producer for a decade or more, continues to turn out its usual complement of gold dust, its entire output amounting now to about $7,000,000, a portion saved by the hydraulic process.

In El Dorado, Amador, Calaveras and Tuolumne counties some drift mining is carried on, Placerville and Mokelumne Hill being the most active centers of this class of mining. Many drift claims were opened years ago under the "table mountain" that traverses Tuolumne county, but the most of these enterprises proved unfortunate and but little has been done there of late years.

Working the River Beds.

The interest displayed in river-bed mining is on the increase year by year and promises to add materially to the future output of the State. River-bed mining consists in diverting the rivers and other large streams wholly or in part from their natural channels, with a view to working the gravel found in their beds. While this business is pursued on most of the larger streams in the mining regions of California, the scenes of the largest operations are the Feather river, in Butte county, and Scott, Salmon and Klamath rivers, in Siskiyou county. This, too, is a revival of another of our primitive methods of mining, having been extensively practiced here in early days. Their beds having been pretty well worked out, many of the streams were abandoned years ago. Afterward there was a general return to the business, it having been found that the beds of these streams had again become enriched through the influx of tailings from the mines being worked along and adjacent to their banks. Lastly, the tailings that were formerly suffered to run to waste are now saved and treated with good and sometimes highly remunerative results. The waste matter from the hydraulic mines, which in many instances has accumulated in great quantities along the outletting channels, is being in various localities subjected to a rewashing and made to yield satisfactory wages. In like manner many of the old ore dumps are being sorted over and cullings reworked, the latter yielding often more metal under the new processes than was obtained from the ore at its first handling.

The object of river-bed mining is to recover the gravel forming the bottoms of river channels or streams and known to be auriferous. To do this various expedients are resorted to, such as draining the channel, wholly or in part, subaqueous armor, dredging, etc. Where it is sought to drain the whole bed of the stream the water is diverted by means of dams into a ditch or flume constructed along the bank of the stream to a point below the section to be reclaimed, and there the entire flow is returned to the channel. By this means such section can be so far freed from water that it is possible to control the seepage by pumps, wheels, etc. Where there exist natural facilities for running tunnels, the entire river bed can in like manner be laid bare by such means. When the design is to dry and work only a strip along one side of the river-bed this is effected by what in mining parlance is termed a "wing dam," that is a water-tight wall which starts from the bank and is carried out a short distance into and down the river, the wall being continued back to the bank. The water inside the space so inclosed is then raised with wheels or hand pumps and emptied into flumes that discharge it into the river.

The above comprise the only methods successfully employed for river-bed working in California. The trials made with dredgers, diving apparatus, etc., have proved failures alike in our river channels and in the gold-bearing sea sands along our northern coast.

While not peculiar to California, river-bed mining has been pursued here on a scale not paralleled in other countries, and the efficiency of our methods greatly surpass those employed elsewhere. Outside this State the business does not appear to have reached large proportions, nor has any great amount of gold been gathered elsewhere by this method.

Working the beds of the rivers that traverse the mining regions of California was begun here at an early day. The first crop of gold dust harvested by this mode, however, was very bountiful. Like some other kinds of gold mining here this branch of the business, after having prospered and attained large dimensions, underwent a marked decline. It has for several years past been on the increase, however, both as regards the number and magnitude of the operations.

The northern tier of counties is distinguished for the many river-bed operations in progress there. The business in that section of the State is prosecuted mostly by the wing-dam system. Many of the claims are worked by the Chinese, who hold some by location, but more by purchase or under lease from the whites. Several thousand Mongolians are engaged in this class of mining in that region. For the time they are at work they make good wages. Their annual earnings are estimated to aggregate a million dollars at has exceeded the above amount, though their average earnings are of course much smaller. These companies are numerous along these northern rivers. Being able to work their claims only during the summer and fall months, this class of miners turn their attention to other pursuits for the rest of the year, such as farming, lumbering, fruit-growing, etc.

The following constitute the principal localities in which river-bed mining is now being carried on elsewhere in the State: Along the several forks of the Yuba, the American and the Feather rivers there are many small Chinese with a few larger white companies engaged in reworking the beds of these streams, the greater portions of which have been gone over and cleaned out many years ago. These river-beds have since become so much enriched through the deposit of tailings from the mines, chiefly the hydraulic washings, that they can, with the present improved gold-saving appliances, be reworked with profit. There are here, too, some spots of virgin ground that, accidentally passed over by the pioneer miners, remain to bless the gleaners of the field.

Most of the operations along these streams, as well as at the few points further south, where any of this sort of work is being done, are carried on either by wing-damming or by diverting the water into artificial conduits along the river banks, freeing the entire channels. Recourse to tunnels for effecting the same end is had in only two or three localities.

Quartz Mining.

So far as productiveness and extent of operations go, however, quartz or vein mining is the leading branch of the business in California, fully two-thirds of the gold product of the State being obtained from auriferous ores. This branch of mining, says an authority, is spread over the entire length and nearly the entire breadth of California, being pursued to some extent in three-fourths of the counties of the State. This industry employs about 4000 stamps or their equivalent, some of the crushing being performed by arrastras, roller mills and similar devices. Of the above number it may be calculated that 3500 stamps are constantly in active service. Estimating that these stamps crush ten tons of ore per day for 300 days in the year, there results an annual total of 2,100,000 tons of ore crushed. As this ore will average nearly $7 per ton, the yield amounts, at the lowest calculation, to $13,000,000 per annum. That this product will be steadily increased for many years to come there is good reason to believe. Nevada, Amador and Sierra remain the leading quartz mining counties of the State, their annual output amounting to $3,000,000,000, $2,000,000 and $1,500,000 respectively. The new impetus in quartz mining is due to the introduction of improved mechanisms, appliances and processes. Through the use of these aids the tendency is constantly toward the working of poorer ores and other low-grade material, so much so that mines not long since considered worthless are now being operated with profit. Gold-bearing quartz is now being milled in this State, and made to pay, that yields a total of less than $2 per ton, the conditions in such cases being, of course, exceptionally favorable. Then, too, invention is ever on the rack to discover new means of reducing rebellious ores, the steady resultant being an ever increasing output of gold. Again, science has been called in and the extraction of gold from sulphurets is no longer a mere mechanical process, but involves wasting, treating with chemical solutions and other intricate and delicate operations known to metallurgists. Many a mine really depends for its success upon the adoption of the most suitable method for dealing with the sulphurets.

There are perhaps 100 arrastras running in different parts of the State, some of them by water, the greater number, however, by horse or mule power. The latter crush an average of one ton, and the former two to three tons per day. These machines are employed where there is only a small amount of ore to be crushed, and which must necessarily be of good grade to justify its being worked by this slow method. The arrastra process is a favorite one with the Mexicans, in whose country it is largely adopted in both gold and silver mining.

Of our California quartz mills, about 60 per cent are run exclusively by water, 30 per cent wholly by steam, and 10 per cent by both water and steam, the latter being used when the water fails, as frequently happens toward the end of the dry season.

Attached to a few of the larger mills are chlorination works for treating the sulphurets saved by concentration, now practiced where the ore carries any considerable percentage of auriferous sulphurets, as most of the California gold-bearing quartz do.

The stamps in use with us range in weight from 400 to 1000 pounds each. The average is about 800 pounds or a little less. In former years they were much lighter than now, the tendency having been steadily toward increased weight. In only a few instances, however, have stamps been used weighing as much as 1000 pounds each. There prevails among our millmen a disposition to find something that will do not only cheaper but better work than the stamps, and many experiments with the various other machines mentioned are being made to that end. That either these or other more highly perfected devices will succeed in largely, if not wholly, supplanting the stamp is not improbable. The latter has, however, succeeded in keeping its place in most of the larger mills.

As in every mining country, the cost of ore extraction and reduction varies over a wide range in California, there being mines in this State where the cost of both operations is reduced to less than $1. These are, however, exceptional cases, nor are they at all numerous, the cost of mining varying here from 40 cents to $3 per ton, and the cost of milling from 39 cents to $2 per ton, the mean cost of the former being about $2 and the latter about $1 per ton. The figures here given refer to our ordinary gold-bearing quartz. There is a class of this ore so debased that the cost of its reduction is much greater than the rates above given. The expense of reducing our argentiferous ores is also greatly in excess of these rates, some of these ores requiring to be treated by roasting or smelting, though generally susceptible of reduction by the simple mill or pan process.

Exclusive of the big establishments designed to buy ores and do custom work, there are not more than a dozen smelters in the State, the most of these being located in Inyo county, only a small portion of the whole being now in operation. The silver stamp mills are included in the list of quartz mills.

Other Kinds of Mining.

Besides those already mentioned and partially described, the gold-bearing deposits of California occur in several other forms, all designated by names more or less fit, a few being perhaps a little fanciful. The most of these deposits are, in fact, distinguished not so much by any inherent peculiarities as by the conditions under which they are found and the methods and appliances adopted in working them.

The auriferous beach sands, which once afforded profitable employment to many men, have years since become so impoverished that they figure no longer among our available mineral resources. These ocean placers have, in fact, responded so feebly to the attempts made of late to work them that beach mining may be ranked among our extinct industries. But, for all this, we have these deposits of low grade in indefinite quantity occurring at intervals. They reach along the seashore for many miles, extending at several points, in the form of buried channels, some distance inland. So abundant, but now so poor, these gold-bearing sands await the coming machine that is to make their further working profitable. Many machines claiming the ability to do this have already been invented and tested, but none of them have fully, or even more than partially, met the requirements of the case.

Meantime the auriferous beaches continue to be worked at a few points and in a small way. Along the seashore in Humboldt and Del Norte counties, formerly the chief sites of this class of mining, the residents of that section of the State gather from these sands by hand sluicing a little gold every year. Their earnings are small and their labor intermittent, being prosecuted only when they have water for the washing, which in most localities is the case during only a small portion of the year.

Besides these "gold bluffs" and "beaches" we have in California a variety of other auriferous deposits, some of which, like the gold bluffs, are peculiar to the State; nor do more than a few of the others meet elsewhere with such large development as here. The principal of these deposits, designating them by the local names, consist of the following, viz.:

The dry diggings, so called, are simply such surface placers as, being without a sufficient natural supply of water for washing, cannot be supplied by artificial means. There are many localities of this character in California. In cases of this kind, if the auriferous earth is not rich enough to bear transportation to water, the gold is separated from it by "dry washing," a process formerly conducted by means of the Mexican batea, still employed in some places. By the Spanish-speaking races the batea continues to be exclusively used in the dry diggings, and these people are very skillful in handling it. Latterly dry-washing machines of various kinds have been invented, some of which are efficient, as much so, in fact, as can reasonably be looked for, considering the inherent difficulty of the work. An entirely satisfactory dry washer remains, however, a desideratum. There are in this State extensive deposits for which the dry washer alone is adapted, but these remain little utilized, owing to lack of a more effective machine of this kind. These deposits occur mostly on the Mojave and Colorado deserts. Some are met with, also, in Los Angeles and San Diego counties. When found farther north they are situated for the most part in small gulches and flats, often at considerable altitudes.

The seam diggings consist of narrow veins of auriferous quartz, varying from not more than half an inch to an inch or two in thickness, found in this State occasionally traversing other formations, and which but for their extreme richness would not justify the expense attendant on extraction. Carrying so much gold as they do, the working of these veins has generally proved remunerative. The weak point about these "razor-blade" veins, as they are called, is their unreliable character; seldom do they extend to any great depth, nor does their width of gold always run with their downward continuity.

The best paying deposits of this kind were found some years ago in Greenwood valley, El Dorado county. They yielded largely for a time, but are now pretty well worked out. In the South Fork district, Shasta county, occur many of these narrow veins, their average thickness being about three inches. They are not so rich, but they go deeper here, more generally than has elsewhere been the case, some of them carrying their usual quantity of gold and holding it for forty or fifty feet before the inclosing granite pinches them out. In this locality the ore taken out is worked in arrastras; there have for many years been five or six of them running in the district, earning for the owners very fair and occasionally large wages. These machines are driven by water and crush from two to three tons of ore per day. As a rule the quartz mined in the "seam diggings" is worked in hand mortars; its small quantity and great richness rendering this the most desirable method for its reduction.

The cement deposits are composed of the indurated gold-bearing gravel taken from the hydraulic and drift mines, mostly from the latter, and which, owing to its hardness, has to be crushed with stamps. This indurated gravel is met with more largely in the southern than in the more northerly drift mines, 75 of the 100 stamps employed in crushing it being in Nevada and Placer counties. As the hydraulic washings approached bedrock more of this material was encountered, and but for the check put on this class of operations twice as many stamps as are in use at present would probably be employed crushing cement.

"Pocket" mining consists in the exploitation of that class of quartz lodes in which the available ore occurs mostly in the form of rich bunches or "pockets." While these bunches are apt to be much scattered, occurring only at long intervals, this is sometimes a lucrative branch of mining. Its grand chances prove very alluring to the more adventurous class of prospectors. While rich pockets have been encountered in the quartz lodes in all parts of the State and throughout the entire history of mining, Tuolumne county has been most distinguished for deposits of this kind. From what is known as the Bonanza claim, near the town of Sonora, there was claimed to have been taken during the four years preceding 1882 nearly $1,000,000, all realized at small expense—not more than half a dozen laborers were employed. Since that time the claim has yielded, it is stated, with equal net profit, about as much more. Since 1852 this neighborhood has been noted for finds of this character. During that year

uilding undermined.
eds of pounds weight
ce straws in the cur-
ains were moved in
y topography of the

note here that while
o ourselves of having
sluicing, our mighty

$2,000,000 to $3,000,000 to the mile.

According to the reports of the engineers detailed by the Government to examine into the question of mining debris, there were some 857,000,000 cubic yards of material excavated during the prevalence of hydraulic operations, of which 230,000,000 yards remained in the beds of the three principal rivers affected—the Yuba, Bear

THE MAN WHO DISCOVERED GOLD.

exaggeration of
omans in Spain.
other labor, too,
e which entails

and American. After a careful examination of the damage done by this debris the engineers reported the following as the injury done along the three streams

pended, contained the following clau
the case being that of Woodruff vs. t
North Bloomfield Mining Company et a

"On consideration whereof it is by
court ordered, adjudged and decreed
follows, to wit: That the defenda
herein and their and each and all of th
servants, agents and employees are p
petually enjoined and restrained f
discharging or dumping into the Yuba
into any of the forks or branches or
any stream tributary to said river or
of its forks, ravines or branches, and
pecially into Deer creek, Sucker Flat
vine, Humbug creek, Scotchman's cr
any of the tailings, bowlders, cob
stones, gravel, sand, clay, debris or re
matter from any of the tracts of min
land or mines described in the compl
and also from causing or suffering to
into said rivers, creeks or tribu
streams aforesaid therefrom any of
tailings, bowlders, cobble-stones, gra
sand, clay, or refuse matter resulting
arising from mining thereon; and
from allowing others to use the w
supply of said several mines or mi
claims or any part thereof for the purp
of washing into said rivers and str
any earth, rock, bowlders, clay, sand
solid material contained in any plac
gravel ground or mine."

As a result of this inhibition a prod
of $10,000,000 annually was cut off, a l
share of which had found its way dir
into the channels of trade. At the
time property in which had been inv
fully $100,000,000 was made useless,
has remained so to this time.

In reply to the question, "Can hydr
mining be resumed without injury t
navigable streams?" the Board of
gineers reported:

"It is not apparent to the board
any expression of opinion or recommen
tion will have any effect in rehabilita
the industry in the present legal statu
the question. Without some modificati
then, of existing conditions hydrau
mining must cease. It cannot be carr
on without violating the decrees of
courts.

"If, however, by a reversal of the o
ions of the courts or by other mean
draulic mining be permitted in who
in part, or if without such reversal an
pression of opinion is required as to t
feasibility of impounding mining debr
the board will state that the investi
tions and examinations made indica
that in isolated cases it is possible to im
pound debris without injury; also, th
locations exist in the canyons of t
different mining streams in the Sierra di
trict where permanent stone dams, pr
perly constructed, will retain large qua
tities of material of the character former
mined out and which caused the destru
tion of the farming lands and injured th
navigation of the rivers.

PLACER MINING.

a party of Mexicans took out on Bald mountain, two miles north of Sonora, as much gold as would load a mule, exactly how much was never known. Near Littleton, a few miles south of Sonora, two miners came upon a nest of these "chispis," and gathered over $100,000 worth. From a claim at Don Pedro's bar, in this county, there was taken some years ago the sum of $100,000, at a cost not to exceed $5000. From the Morgan quartz claim, on Carson hill, just over the line in Calaveras county, there was, in the early fifties, pounded out with a hand mortar and pestle, gold valued at $3,000,000. With such results extending through so many years and scattered all over the State, it is not strange that this exploiting for pockets should be with many a favorite style of mining.

Hunting for "nuggets" is carried on in both vein and placer deposits. The greatest success of late has been met with in the latter. During the year 1889 Appel & Grant, working their quartz claim at Chip's Flat, Sierra county, it is recorded, took out in a few months, and with little more cost than their own labor, over $100,000 worth of nuggets, besides large quantities of rich ore not yet reduced. From the Baughart mine, located twelve miles northwest from the town of Shasta, there was taken, several years since, a large number of nuggets which weighed over a pound each, besides many of lesser weight. From a placer claim situated on the Monte Cristo gravel range there was taken a lot of nuggets ranging in value from $300 to $800 each. These nuggets much resembled in size and form small cobble stones.

Year	California.	Other States.	Total Product.
1848	$10,000,000		$10,000,000
1849	40,000,000		40,000,000
1850	50,000,000		50,000,000
1851	55,000,000		55,000,000
1852	60,000,000		60,000,000
1853	65,000,000		65,000,000
1854	60,000,000		60,000,000
1855	55,000,000		55,000,000
1856	55,000,000		55,000,000
1857	55,000,000		55,000,000
1858	50,000,000		50,000,000
1859	50,000,000		50,000,000
1860	45,000,000	$1,000,000	46,000,000
1861	40,000,000	3,000,000	43,000,000
1862	34,700,000	4,500,000	39,200,000
1863	30,000,000	10,000,000	40,000,000
1864	26,000,000	19,500,000	46,100,000
1865	28,500,000	24,725,000	53,225,000
1866	25,500,000	28,000,000	53,500,000
1867	25,000,000	26,725,000	51,725,000
1868	22,000,000	26,000,000	48,000,000
1869	22,500,000	27,000,000	49,500,000
1870	25,000,000	25,000,000	50,000,000
1871	20,000,000	23,500,000	43,500,000
1872	19,000,000	17,000,000	36,000,000
1873	17,000,000	19,000,000	36,000,000
1874	18,000,000	15,400,000	33,400,000
1875	17,000,000	16,400,000	33,400,000
1876	17,800,000	22,100,000	39,900,000
1877	15,000,000	31,800,000	46,800,000
1878	15,300,000	35,900,000	51,200,000
1879	17,600,000	21,200,000	38,800,000
1880	17,500,000	18,500,000	36,000,000
1881	18,200,000	13,500,000	34,700,000
1882	16,800,000	15,700,000	32,500,000
1883	14,120,000	15,500,000	30,000,000
1884	13,600,000	17,200,000	30,800,000
1885	12,300,000	14,100,000	26,400,000
1886	13,200,000	16,400,000	29,600,000
1887	11,800,000	20,700,000	32,500,000
1888	10,100,000	19,900,000	50,000,000
1889	10,800,000	22,220,000	32,500,000
1890	9,900,000	21,900,000	31,800,000
1891	10,400,000	21,300,000	31,700,000

Gold Not Exhausted.

In corroboration of the position assumed at the outset that the gold deposits of this State are not by any means exhausted, the opinion of the State Mineralogist, William Irelan Jr., may be cited, together with interesting statements bearing upon the subject of mining in general, some of which have already been quoted.

The impression widely obtains, he says, that the gold mines in California have been depleted below the point of profitable production. Many otherwise well informed persons entertain this idea. Nothing can be more erroneous. The gold taken out has exhausted but little of our auriferous wealth, nor has the annual production heretofore much exceeded what we may reasonably hope to reach and maintain in the future.

Again, it is a mistake to suppose, as many do, that the earnings of the pioneer miners were greatly in excess of those at the present day. They were, to be sure, somewhat larger, but not in the proportion popularly believed.

During the era of the largest gold production in this State—say from 1850 to 1855 inclusive—the annual output of gold averaged about only $55,000,000. As the mining population numbered, meantime, about 150,000, their individual earnings averaged barely $366 per year, not much more than the smaller population now engaged in the mines are able to earn, working by no means so many days in the year as their predecessors. Those who work for wages do nearly as well now as they ever did, all things considered. But it now requires a larger amount of both skill and capital to accomplish much in our mines than were needed in the early days, a condition of things that puts the mere wage-earner and worker at a disadvantage.

As regards the extent of our mining field it is simply illimitable. A hundred millions of additional capital might as well be invested there as not, nor would 100,000 men crowd it any more than 60,000. Of the mineral deposits that actually exist in California not a tithe probably has yet been discovered, nor has a much larger proportion of those already discovered been developed to a productive condition. We have made a good beginning—hardly more.

Mining in this State is not now confined, as formerly, to the production of the precious metals. While gold mining continues with us the leading branch of the business, several of the inferior metals, as well as many of the useful minerals, are now produced here in considerable quantities. Of the latter there remains still a number with which little or nothing has been yet done, though we have them of good quality and in the greatest abundance.

Besides her gold fields, the most extensive and prolific of any in the world, and silver-bearing lodes in countless numbers, California possesses the more common metals and minerals in great variety. This State is amply supplied with deposits of iron, tin, lead, copper and quicksilver; borax, salt and soda; petroleum, natural gas and asphaltum; gypsum, steatite, graphite, manganese and chromium, and with coal, nickel, antimony, asbestos, cements, ochre, sulphur and magnesia to a more limited extent; the plastic clays, infusorial earth, lime and building stones, including the fissile slates, abounding in many parts of the State.

There is scarcely a county in California but possesses valuable mineral deposits of one kind or another, the wide distribution of these products being something remarkable. Of the fifty-four counties in the State fourteen make a notable production of gold and twelve of both gold and silver, there being a number of counties in which these metals in smaller quantities are turned out every year. Five counties produce more or less quicksilver, two borax, three salt, four asphaltum, two petroleum, three copper, etc.

Were California even poor in the precious metals she would yet become a great mining State. With such wealth as this she is, in this respect, destined to be a very important factor in the financial affairs of the world.

Gold mining is with us in an embryonic state. It has not yet reached even the stage of sturdy infancy. Our true golden era rests in the future, not in the past. Our El Dorado has not yet been revealed to us. It lies buried deep in the bowels of the earth.

The placer deposits that have made for us such a name and given to mining such impetus and eclat were but driblets which nature, having released from their matrices, brought within our easy reach as a means of encouraging us to further efforts, and leading us on to that greater and more enduring wealth stored away in the rocky ribs of the mountains.

What is here claimed for the future of mining in California is strongly foreshadowed by what has already taken place. For several years past our annual output of bullion has been considerable, and but for the suppression of hydraulic mining, formerly a prolific source of production, would have shown a marked increase. That gravel washing by this method will be resumed, at least in part, we have reason to hope. It would hardly be creditable to our engineering skill should we fail to devise means and methods whereby this class of debris could be so disposed of that hydraulic operations might be largely carried on without serious detriment to other interests. Could this very desirable end be reached the gold product of the State would at once be advanced by several millions annually.

There has among writers on the subject ever existed a wide difference of opinion as to the number of men engaged in the business of mining in California, and for that matter in other of the Pacific States and Territories. H. C. Burchard, Director of the Mint, in his report for 1882, estimated the number throughout our entire mining region as follows:

Arizona.......... 4,678|Nevada.......... 6,674
California......37,147|New Mexico.... 1,496
Colorado........28,970|Oregon.......... 3,696
Dakota.......... 3,570|Wyoming........ 328
Idaho............ 4,708|Utah............ 2,592
Montana........ 4,813|
Total............................98,672

This, though much larger than the number fixed on by some, was at the time probably very nearly correct, the total in all these States and Territories, except California and Nevada, having been somewhat, and in most cases, largely increased since, reaching now 140,000 at least. By this is meant persons engaged directly and indirectly in mining for gold, silver, lead and copper, there being a good many in California, with a few also in some of the other States and Territories named, engaged in various other branches of mining.

In apportioning the present mining population among the different States and Territories they may be assigned as follows:

Arizona.......... 5,000|New Mexico.... 6,000
California......37,000|Oregon.......... 5,000
Colorado....... 35,000|Wyoming........ 1,000
Dakota.......... 5,000|Utah............ 6,000
Montana........25,000|
Idaho...........15,000| Total..........146,000
Nevada......... 6,000|

In the absence of any official count there can only be claimed for these figures an approximate correctness.

It is more difficult to arrive at accuracy on this point in California than in our other mining States and Territories, owing to the much larger number of self-employers we have here, the many different kinds of mining in which they are engaged and the manner in which they are scattered over a great extent of territory. In these other countries mining operations are carried on more by large companies, the number of whose employes can easily and definitely be ascertained.

Where men are employed by scores, hundreds and even thousands, as on the Comstock range and in the big mines of Utah, Colorado and Montana, it is much less troublesome to take their census than where a like number is scattered along the "dead" and the "live" rivers, the gulches and ravines or throughout the hydraulic, drift and quartz mines of California. Dispersed over such wide area and hid away in the deep gorges and canyons or toiling in dark pits and tunnels, a good many of these miners would be missed were even a careful enumeration of them undertaken.

Of the entire number of California miners some 10,000 or 12,000 consist of Chinese, about one-third of whom are employed by the whites on wages, the balance working on their own account or for companies composed of their own countrymen. Very few of this race engage in vein mining on their own account, nor are more than a few of them so employed by the whites, as they have a great aversion to deep underground workings. They confine themselves mainly to the various branches of placer mining, such as working over the partially exhausted or wholly abandoned bars and gulches, reworking tailings, and in river-bed operations, the former effected by hand-sluicing, the rocker also being sometimes employed, and the latter mainly by means of wing-damming.

The Chinese engage in but little drift mining, never in a large way, though some of them are employed by the white drifters, and also a few about the quartz mills. In the northern counties the Chinese carry on hydraulic washing in a few places. In some cases they own the ground, though oftener it is owned by the whites, being worked under lease or on shares. The enactment of the Exclusion law has tended to draw this class of foreigners away from the mineral districts, the increased demand for their services elsewhere insuring the most of them better wages than they can earn in the mines.

In California the miners, more especially those engaged in placer operations, not only are self-employers to a much greater extent than is the case elsewhere, but their labors are here largely intermittent. Very few of them, except those engaged in vein or drift mining, work steadily throughout the year. The river-bed miner can prosecute his labors only in the dry season when the streams are low; other classes work to better advantage in the wet season, when the water is plentiful. Only then can the so-called "dry diggings" be worked or ground sluicing be carried on. Much of the hydraulic washing is also confined to this season.

Owing to this condition of things, most of our placer miners devote a portion of their time to other pursuits, such as farming, fruit and stock raising, lumbering, etc. They are apt to be land-owners in a small way, nearly all of them possessing an orchard and garden, with a few acres for grain growing and pasturage. The major portion of the gold fields are admirably adapted for grape and fruit culture.

The wages paid underground miners in this State, both vein and drift, are almost uniformly $3 per day. Millmen and other above-ground hands receive from $2 50 to $2 75 per day. Chinamen, where employed, are paid about one-half these rates. This is without board and lodging; when these are included the miner is charged for them at the rate of about $6 per week.

If the average earnings of the miners who work their own claims fall below the above rates it is to be considered how much of their time is to be given up to other pursuits which, besides contributing largely towards their livelihood, insure them always comfortable homes. In no other part of the world does the miner live so well nor is he so independent as in California. In comparing the annual bullion product of this with that of other countries the above consideration should also be given due weight.

Besides our output of bullion we produce here of the economic minerals and metals values to the amount of several millions annually; far more than is produced by any one of our neighbors or perhaps by all of them put together.

These several industries, omitting the less important, give employment to some 4000 men, distributed about as follows: Quicksilver, 1000; borax, 300; salt, 400; petroleum and natural gas, asphaltum, 800; coal, 250; chromium and antimony, each 50; slate, marble and other stone quarries, 500; lime, gypsum and the plastic clays, soda, tin, manganese, etc., 1000.

WHERE GOLD IS FOUND.

It Exists in Practically Every County in the State.

In order to show the widespread presence of gold throughout the State each county will be taken up briefly and the leading mines referred to. It will be seen that with one or two minor exceptions gold exists in paying quantities in every county in the State, from San Diego on the south to Siskiyou on the north, and from Alpine on the east even into the sands of the Pacific ocean on the west.

The accompanying statements are not mere hearsay or idle rumor, but are derived from the report of experts in the employ of the State Mining Bureau, and are conservative to a degree, being reliable to the fullest extent. The showing will be as surprising to many Californians, as it undoubtedly will be to those who are residents of other States.

ALPINE.

The lower middle portion of this county abounds with gold and silver bearing lodes, a majority of them being of regular formation and large dimensions. Most of the ores, however, are of low grade and more or less base, making their reduction difficult and expensive. The facilities in the way of wood, water, etc., are such, however, that with proper management and the use of the latest improved methods there is no reason why these deposits should not be profitably exploited. It should be mentioned, by the way, that the first copper deposit ever found in California is located in this county, being known as "Uncle Billy Rogers'" copper mine. It is situated in Hope valley in the northwestern part of the county, and its discovery antedates the finding of the Comstock lode by several years. There are several districts in which much work has been done in the past, including the Monitor, Mogul, Silver Mountain, Silver King, Hope Valley and Blue Lakes.

AMADOR.

This county stands in the very front rank of the bullion producing sections of the State. It is crossed by the mother lode, and along that notable ledge are some twenty-five quartz mills in active operation, with upward of 650 stamps at work. These are all included within a belt about fifteen miles in length. The mother lode between Plymouth and the Mokelumne river is covered by United States patents. Near Plymouth there are claims being prospected. Near Drytown a number of mines have formerly been worked more or less which are now idle, as for instance the Potosi, Italian, Seaton, and the North Gored; near Amador City, the King, the Little Amador, the South Keystone, Median and El Dorado; near Sutter creek (where was situated the famous Hayward's Eureka) undeveloped properties are numerous—the North Lincoln, the Occident, the Comet and Wabash and Mechanics' mine. Just south of the Eureka, the Summit is situated; and on Kennedy flat the Clyde, the Volunteer and the Pioneer immediately adjoining the Kennedy on the south, and partially prospected by its shaft come the Hoffman & Bright properties; the Doyle in Hunt's gulch; the Valparaiso near Middle bar—connected with which is a Huntington mill in operation; the New York claims; the old Hardenburg mine and McKinney properties and many others. Among the bullion producers are the Keystone Consolidated, South Spring Hill, Talisman, El Dorado, North Star, Lincoln, Pioneer, Wildman, Amador Consolidated, Amador, Kennedy, Summit, Zeile, Bell Wether, Volunteer, McKinney & Crannis, Hardenburg, Sargent, Oneida, Plymouth Consolidated, Gover, Bunker Hill, Kennedy, Sutter Creek, Mammoth and many others.

BUTTE.

The gold mines of Butte were among the most famous in those early days that have now become only a tradition, and millions of dollars were taken from her gravel and quartz deposits. While the cessation of hydraulic mining shut off a large part of the production of gold, nevertheless much continues to be done, and at the present time every kind of mining is carried on with success.

The most extensive river-turning enterprise ever undertaken is now in progress near Oroville, the entire flow of the Feather river having been lifted from its course by means of a dam and system of flumes, thus laying the bed of the stream bare for washing.

The Big Bend tunnel, constructed for draining the bed of the Feather river, is not only the largest enterprise of the kind in California, but the largest probably ever undertaken for a similar purpose. The operations of the Spring Valley Hydraulic Company at Cherokee in this county are also among the largest now carried on in the State. In this locality, too, was picked up a majority of the more valuable diamonds found in California. In Butte the pliocene river system, the principal site of the drift mines, meets with its greatest development. Of drift gravel mines the Bay State, Eureka, Magalia Consolidated, Oro Fino, Lucretia, Aurora and Indian Spring are among the most prominent. The principal quartz mining localities are Forbestown, Wyandotte, Cherokee, Brown's valley, Merrimac, Yankee Hill, Inskip and Oregon City. There are eleven quartz mills in the county, and the present season has witnessed a decided revival in mining operations. The quartz deposits already known to exist afford opportunities that are practically inexhaustible.

CALAVERAS.

The very name of this county brings to mind the memories of the old mining days, but if any one fancies for a moment that the gold deposits of Calaveras are exhausted he is sadly mistaken. Mining is still carried on here in nearly all its forms, and there are mines at Angels, Murphy's, Copperopolis, Milton, Mokelumne Hill, Campo Seco, Sheep Ranch, Altaville, West Point, Rich Gulch, Douglass Flat and elsewhere which are still paying handsomely. The Utica mine at Angels is the leading quartz deposit of the county. The Stickies at the same place is another good mine, and so are the Smyth and McCreight mines in the vicinity of Angels. Other prominent mines are the Sheep Ranch, Esmeralda, Ilex, Torchwood, Blazing Star and Water Lily, Angels, Quaker, Buena Vista, Union, Plymouth Rock, Calaveras, etc. There are chlorination works at Angels and

West Point which carry on a profitable business in the reduction of sulphurets. It is certain that there are still extensive gravel deposits in this county which would pay well for working, while there is quartz enough to keep all the mills in the county busy for a century.

COLUSA.

Colusa, though a great wheat-growing county, contains a variety of mineral products, the more important of which consist of gold, copper, cinnabar, sulphur, coal, petroleum, bitumen, natural gas, clay and limestone. There are also several mineral springs in the county, some of them noted for their medicinal properties. The principal gold quartz mines in the county are the Clyde and the Manzanita, located on Sulphur creek. Along Bear creek and in the vicinity of these mines many of the gulches have afforded some placer diggings, but being neither rich nor extensive the amount of gold obtained from them has not been large. The Manzanita mine is located on Sulphur creek, twenty-seven miles southwest of Williams, a town on the California and Oregon railroad. The original claim, located February 21, 1863, is 2700 by 1000 feet. The country rock here consists of sedimentary shales and a sandstone with occasional outbursts of eruptive rock. In many places these rocks are coated with a siliceous sinter, evidently deposited from hot siliceous waters, traces of which, in the form of thermal springs, are still plainly visible. The substance not only coats the rocks, but it has found its way into all their cracks and crevices. It has, as a rule, free gold associated with it and constitutes the auriferous ore of the district. The gold does not appear to permeate the quartz, but is deposited on it in the form of an incrustation. This is the case at least in such parts of the mine as contain much of this sinter. In many places this sinter is associated also with cinnabar and bitumen, which latter is often in such quantity that it causes great loss of gold, imparting a coating to the water that causes the gold to float off with the tailings.

CONTRA COSTA.

Contra Costa county is known chiefly from a mineral standpoint as the center of the coal-producing region of California, and this feature of her resources will be found dealt with elsewhere. But in addition gold and silver have both been found, though not in any appreciable quantities. Still the precious metals are known to exist here, and doubtless close research would develop one or both in paying amounts.

DEL NORTE.

This county was the scene of one of the famous gold rushes in the early period of the history of the State. The streams in the interior carry considerable gold-bearing gravel, while there are deposits of black sand at different points along the coast which are known to contain much gold. Some day a process will be discovered for working these sands successfully, and when this shall have been done a great and profitable industry will result. Some placer and hydraulic mining is still carried on along Smith river and the tributaries of the Klamath, and an occasional deposit of quartz has been found. Just over the line in Oregon from Del Norte are large gravel deposits which have paid handsomely in the past, and which are known to be rich in gold.

EL DORADO.

This is the fitting name of the county, where the discovery was made which literally revolutionized the finances of the world. It was at Coloma that Marshall, made the memorable discovery on that January morning in 1848, and on the hill, just above that town stands the monument which California has erected in commemoration of the event which meant so much to her. It must not be supposed that the mineral wealth of this section has been exhausted simply because the halcyon days of the rocker and pan have long since passed away. On the contrary, the development of the vast quartz deposits has scarcely been commenced, and in these hills lie treasures far surpassing in amount all that have yet been wrenched from their grasp. A broad belt of quartz veins extends through the county which has been tapped here and there, but which presents the most favorable opportunities for capital whenever the popular idea in regard to the gold mines shall have been revised. There exist large deposits of gravel in various places which can be worked by the hydraulic process, and doubtless will be opened again before long. The mother lode crosses this county from north to south a distance of twenty miles. After entering the county and proceeding a few miles north it makes a rather violent deflection to the east, carrying it into the neighborhood of Placerville. A little farther on it comes back to its normal course, which it holds till it reaches the middle fork of the American river, the northern boundary of the county. There are located along this section of the mother lode, its porphyritic appendages included, nearly a hundred mining claims, on all of which more or less exploratory work has been done, many of these claims having been equipped with costly plant and developed into largely productive mines. Among the leading properties are the Josephine, Oakland, Central, El Dorado, Equator, Superior, Big Sandy, the group at Placerville, the Dalmatia, Esperanza, Ivanhoe, Taylor, Bona Torsa, Zentgraft and others. On the Georgetown divide are many gravel and quartz deposits which will repay working. Much of the ore found in El Dorado county is low grade and many mines were opened in the early days which were abandoned because they did not come up to the exalted ideas of those times. Now that ores can be worked for less than a dollar a ton, as is actually being done at present in El Dorado county, there is no reason except lack of enterprise why these low-grade deposits should not be extensively developed.

FRESNO.

From the earliest history of the State this section has been known to be rich in mineral wealth. Placer mining was begun about the year 1850 in the San Joaquin river and its tributaries, and a mining population of about 1500 sprung up in that vicinity. Many men made small fortunes, while the degree of success of others was less. One of the claims there paid its proprietor $117,000. There were also several Chinese companies engaged in mining who faied very well. At Fine Gold creek alone there were at one time in the early fifties 500 people mining. The placers were worked for all they were worth until the winter of 1867-68, when the heavy floods swept everything away, discouraging the miners, the greater number of whom sought more promising and newer fields. The advent of the agriculturist about this time and the exhaustion of the old placers also contributed to bring about the cessation of placer mining. There are still some very good claims there, but they have never been worked on account of the difficulty of turning the water. Some quartz mining was done early in the fifties. The quartz mining belt is from eight to ten miles in width, running parallel with the Sierra Nevada, its edge being about twenty-five miles from Fresno. The various mining districts are Hildreth, Auberry, Mount Raymond, Fresno Flats, Grub Gulch, Coarse Gold, Fine Gold, North Fork and the Minarets. Many fine prospects have been made, which will no doubt prove valuable properties when developed. When experienced mining men become interested and invest capital—the scarcity of which up to the present has prevented the full development of the mines—the results will be most gratifying. The ores extracted have in many cases assayed very highly, as much as $500 and in some cases even $1000. Of course, this is much above the average, but the ore is of a high grade generally. The capital invested in the mines in this county amounts to fully $350,000, while the number of men employed is about 500. Though many of the mines have been worked quite steadily, the operations, except in a few instances, have not been on a very extensive scale.

HUMBOLDT.

Humboldt county possesses some valuable gold mines, notable being those at what is known as Gold Bluff, where there are extensive black sand deposits. These deposits created much excitement some thirty-five or thirty-six years ago. They extend from Oregon eight miles in a southerly direction. The whole line of the beach was worked for the gold the sand contained. This deposit of black sand contains traces of platinum, osmium —iridium too minute to have any commercial value, but with sufficient gold to make the business of gathering it lucrative. Several parties engaged in this auriferous harvest have retired with a competency. From the date of discovery to the present writing these beaches have been annually worked, success depending more or less, on the occurrence of fierce gales of wind in the winter time, and the consequently heavy surf that breaks along the shore. These black sands, as they are called, come from the disintegration of the bluffs facing the ocean, and which rise nearly vertical to a height of from 100 to 800 feet above the sea level. The bluffs are a formation of auriferous gravels deposited by the Klamath and its tributaries, the main river emptying into the sea centuries ago at this point. The bank caves, and the retreating waves carry the lighter detritus seaward, while the metallic portion remains on the beach in thin sheets, which the miner gathers and washes. This gold is from 900 to 950 fine, and sells for $19.50 per ounce. It is estimated that over $1,000,000 have been taken from this source with comparatively small expense. Besides these beach deposits there are some fifteen hydraulic mines on the Klamath and its tributaries which have been or are now successfully worked. The inhibition against hydraulicking does not apply in this section, hence the non-cessation of the industry.

INYO.

This county has extensive deposits of gold, silver and other minerals, and in the past has produced a large amount of bullion. The names of the Panamint, Darwin and Cerro Gordo camps are well known as having at one time been among the most prominent mining localities on the coast. The remoteness of these places and their difficulty of access has prevented rapid development, though there are extensive deposits of ore which will repay working.

The miners all along the Inyo range for 150 miles north either work in their own claims and sell their small batches of assorted ores or work on tribute, either of which secures them living wages.

The Lookout or Darwin district adjoins the Panamint district on the northwest and takes its name from the principal town in it, situated on the eastern slope of Lookout mountain. There are three smelters in this district, but they have not been run for several years. The most largely developed and productive mines in the neighborhood of Darwin are the Defiance, Independence, Promontory, Sterling, Pluto, Christmas Gift and the Lucky Jim, the last three included in the Mackenzie group, being situated four miles north of the town. The mill was erected on the spot, but, although a tunnel 1200 feet long was run into the hill in hope of striking the ledge from which the bowlder came, nothing remunerative was developed. Various attempts have been made to discover the source of the placer gold, but heretofore prospecting has met with but little success in the county, the great obstacle being the depth of soil covering the rock formation and the dense growth it maintains, both of which prove a hindrance to the prospector and geological observer, yet some few ledges have been unearthed taken up, but little of much value has been developed. Both the mill and smelter that were put up several years ago—the one near the town, the other at Swansea—have been idle for several years; the great cost of fuel and the opportunity of shipping the ores by railroad causing their destruction.

Independence, the county seat of Inyo, is the center of a good many mines, lying in almost every direction around it, some of them being out of the limits of any organized district. Conspicuous among this class is the Brown Monster, located six miles south of Independence station, on the Carson and Colorado railroad. The mine is connected with the railroad by a tramway, also the mill, standing on the bank of Owens river. This mill carries thirty stamps and has a crushing capacity of forty-five tons of gold ore per day.

KERN.

Among the many sources of wealth of which Kern county boasts and which is bringing and has brought hundreds of thrifty emigrants within its borders to live, that furnished by its mineral deposits has been, perhaps, most neglected, by those who have written most of this section, at all events. In 1854 gold was first discovered in Kern county by a party of emigrants while camping in a gulch in the Greenhorn mountains. There was immediately a rush of prospectors to this country, and the Kern river excitement became as noted as that of Fraser river or White Pine.

In April of the same year Captain Maltby discovered the first quartz vein near what is now called the Hot Spring valley. He erected a quartz mill and operated very successfully for two years. In the meantime the bars along Kern river were extensively worked for placer gold, with profitable results.

In 1855, in a cove nestling at the foot of the Greenhorn mountains, Richard Keys and Jonathan Crandall discovered what is called the Keys mine. It proved a rich strike, the quartz yielding about $300 per ton in gold. It is estimated that some $2,000,000 in profits were extracted from the group of mines.

In 1859 Lovely Rogers and Joseph Caldwell found the "Cove" mines near the present site of Kernville, consisting of the mines now known as the Lady Bell, Jeff Davis and Beauregard, all afterward called the Summer mines, and extensively worked by Senator John P. Jones and others. Millions in money were taken out of these mines.

In 1862 the rich mines of Havilah were discovered, and for a season there was a splendid output. What was known as the McKeadney group of mines yielded thousands of tons of quartz paying from $200 to $300 per ton.

The mines in operation at present in Kern county are as follows: Stater & McKay's group in the Agua Caliente region yielding $60 and upward per ton; John's mine at Agua Caliente; Mann's mines, which always produce rich ore; D. Applegate and others in Kelsoe valley and in the vicinity of the noted St. John's mines; W. J. Graham, W. Menzel, the Shipsey Brothers and others in the Hayes valley; the Robinson mine, producing from $70 to $100 in gold per ton; the Melville group of mines and the original Mammoth mine.

Along upper Kern river rich specimens of float quartz have been discovered from time to time, and in that region there is a fine field for prospectors.

That part of the Mojave desert which lies in Kern county has not yet been carefully prospected, yet there exists possibility there of another Calico district, especially in the vicinity of Red Rock canyon.

In general it may be said that the gold mining interests of Kern county have languished for some years, but there are at present signs of renewed life, and there seems to be no reason why this industry should not yield a golden harvest.

LAKE.

Lake county contains a great variety of minerals, gold, silver, copper, borax, sulphur, asbestos and cinnabar counting among the mineral resources. In Paradise valley, about five miles from the sulphur banks, a shaft has been sunk to a depth of sixty feet on a ledge of quartzite. The ore, which is much copper stained, carries considerable pyrites, and assays from $3 to $9 in gold a ton, with a small percentage of silver. Gold-bearing quartz has been observed in the vicinity of Mount St. Helena, also near the Bradford quicksilver mine, and at a point between Anderson springs and the Geysers. The croppings of these quartz veins contain a small amount of silver. One mile east of Bradford much copper float is to be seen and near Harbin springs a shaft has been sunk to a depth of sixty feet in a cupriferous vein, but the ore is of too low a grade to warrant further sinking.

LASSEN.

Lassen county, bordering on one of the best mining counties of California, being separated from Plumas county by a spur of the Sierra Nevada, has so far developed but little mineral wealth, a few claims having been prospected on Diamond mountain, near Susanville, the county seat, that have yielded some gold. Veins of silver and gold ores have also been found on the southwest side of Eagle lake, but mining as a regular business has only been prosecuted in the extreme north of the county, sixty miles north of Susanville and nine miles from the Modoc county line, in what is known as the Hayden Hill Mining District. This hill, named after one of the first locators, who is buried there, is one of the highest points of a spur running out on the Eastern slope of the Sierra Nevada; its altitude is given as 7500 feet. The mines were discovered nearly twenty years ago, since which time they have been more or less continuously worked, yielding to the world's gold supply a little over $1,000,000.

washing of the sea, tically inexhaustible. ATEO. is county, as far as wn, consists of gold, al quicksilver, lime Of these petroleum are at present alone al account.

wheat; there was no fine gold. The gold was worth $18 per ounce. Parties have mined on the creek for five years, and it is stated that over $28,000 worth of gold has been shipped through Wells, Fargo & Co. at Santa Cruz. In early days a bowlder, the size of which was estimated at some sixteen cubic feet, was discovered in Gold Gulch, near Felton, which, when

Another mine which has attracted c siderable attention is the tellurium dep discovered a few years ago some tl miles from Redding in a gulch tha tributary to the Sacramento river. T mine was discovered by Peter Sche who was led to search for a quartz le because of the rich placers that had b found in the vicinity. The first shot

LING GOLD-BEARING GRAVEL.

been discovered in | milled. yielded some $33,000. A small | into the ledge when it was uncovered

FORTY-STAMP QUARTZ MILL.

LOS ANGELES.

The mineral and metallic productions of the county include gold, silver, copper, asphaltum, petroleum, graphite, iron, limestone, gypsum, borate of lime, magnesia, kaolin, borax, alum, salt, building stones such as granite, sandstone, marble, etc. In entering Los Angeles county by railroad from the north, we have on the east side the Mojave desert, an arid scope of country covering between fifteen and twenty townships in the northeastern part of the county. This desert country has been prospected over to some extent and reports are current of the presence of rich base silver ores and also copper; further, deposits of clay and gypsum have been found a few miles back of Alpine, one of the stations on the Southern Pacific Railroad. The prospecting and working of these metalliferous veins is accompanied with so much hardship and expense, on account of the scarcity of wood and water and feed, that the mines have never been developed sufficiently to permit of any decided opinion being formed as to their true value.

The gold district of Los Angeles county is Acton, located about fifty-five miles from Los Angeles, on the line of the Southern Pacific Railroad. There are several producing gold mines in this camp, the most prominent of which are the Red Rover and the New York mines.

On Mount Gleason, eight miles southwest of Acton, some very promising gold prospects are being developed and machinery put in to thoroughly test the property.

In the vicinity of Newhall are placers that have been worked since the first discovery of gold here ten years or more before Marshall's famous find at Colusa. Among these are the San Feliciana diggings, which are situated at an elevation of 2100 feet, between Castaca diggings and Piru creek, twelve mile northwest of Newhall, on the Southern Pacific railroad. This deposit of gravel, for an area of eight by four miles in extent, is supposed to average fifteen feet in depth, and is cut through by gulches and canyons. Each canyon throughout this area has been more or less worked for the last ten years.

During the period from 1810 to 1840 Jose Bermudes and Francisco Lopez superintended the Mission Indians in working these gravel deposits. In 1842, finding that these deposits, though worked in a crude manner, paid exceedingly well, the Mexican Government was petitioned to consider the territory between Piru creek and the Soledad Canyon, and extending west to the Mojave desert, mineral land, and that no grant be extended taking in that territory. This petition was granted by the Government. The most extensive mining operations carried on in this belt of gravel were in 1854, when Francisco Garcia took out of the San Feliciana gulch in one season $65,000 in gold.

MARIN.

This has never been regarded as a mineral producing county, yet it unquestionably contains deposits that will some day prove of value. On the ocean side of Tomales Point occurs a deposit of auriferous black sand. It can be reached only at low tide and has not proved rich enough to warrant continuous working. Besides, the supply is uncertain, being dependent on the winds, the surf and the tides. A little to the east of Tomales some locations have been made on a quartz ledge carrying gold. On the westerly slope of Tamalpais some silver prospects have been found, but the find has not been followed up by any developments.

MARIPOSA.

Gold mining has from the first been in this county its most prominent industry. The placer diggings, though not so extensive as in some other parts of the State, being rich and shallow and therefore easily worked, paid large wages in pioneer times. These surface deposits becoming speedily depleted, early recourse was had here to quartz mining, this industry having been inaugurated on the Fremont estate in 1851. In Mariposa that remarkable auriferous belt known as the mother lode of California has its southerly beginning, it being here displayed in great power. This lode, which strikes nearly north and south, dips to the eastward at an angle varying from 45 degrees to 70 degrees. The walls are uniform; the eastern hanging-wall is greenstone and the western foot-wall is slate. It is to be regretted that the mines belonging to the Fremont estate are not being worked at present, as they have not been for many years past, covering as they do a large portion of the mineral section of the county. This inaction on the part of the owners of the estate has had a depressing effect on the general business of the county.

Mining men have too generally been led into the error of supposing that as the Mariposa grant was for many years involved in ruinous litigation all the mining capabilities of Mariposa county were thereby necessarily in a state of hopeless entanglement. The fact is that the grant only covers comparatively a small portion of the really rich mining territory of Mariposa county, and there are outside of its limits plenty of valuable opportunities for those who are seeking mining ventures.

The whole of the grant is intersected by a network of veins, but very few of which have been opened up to the present time. The region in which the estate lies was noted in an early day for the extent and richness of its placer mines, which still yield good returns when worked during the rainy season.

The general course of the veins is from northeast to southwest, extending through the property in its greatest length. The most extensive explorations have been made on the Princeton, situated near the center of the estate, and which has been traced for three miles and a quarter, and the Josephine and Pine Tree, situated in the northern extremity, near the Merced river, and which seem to be prongs of the same vein.

These are all situated on the great mother lode, which runs through the estate for a distance of ten miles, and the latter veins crop out boldly on the sides of Mount Bullion, which forms a part of the eastern boundary of the property. Some other veins, as the Mariposa and the New Britain, situated near the county seat, Mariposa, and others, have been opened to some extent.

The Princeton has a record of yielding $3,000,000 from workings down to 500 feet in depth. Its ores and general features are very similar to those of the mother lode as seen in Amador county mines.

MENDOCINO.

Mendocino has gained no distinction as a mining county, yet mineral resources are not wanting within her boundaries. Both lode and placer auriferous deposits have been discovered here at different times and in various places. A few of these deposits have been worked in a limited way. Copper ore, some of it promising, has been found in Coyote, Potter and Walker valleys. Exudation of petroleum has also been noticed in several parts of the county. At Punta Arenas this substance trickles from a sandy shale on the seashore. During the oil excitement in California in 1865 quite a sum of money was expended at this place to obtain a more ample supply from the shale, but without success. Sulphur and salt are met with in quantity, and mineral springs, hot and cold, issue from the earth.

MERCED.

Although Merced is one of the foremost grain, wool and fruit growing counties in the State, it is not entirely destitute of mineral wealth. There was at one time in the northeastern corner of the county a limited extent of placer diggings, and both quicksilver and antimony are found in the eastern portion of the McLeod mining district, which extends from San Benito into the southwestern corner of Merced county. Quicksilver is both mined and reduced in that district.

Gold can be found in the sands of all the streams upon the eastern side of the county, and in some places mining is still carried on where the Merced river leaves the foothills throughout a great portion of the year, usually from August until the miners are driven out by the high water. Both white men and Chinese, frequently to the number of about 100, engage in this work, and it is said they make good wages.

There is a bluff on the eastern boundary of the county, about half a mile north from the Merced river, which pays well during the wettest portion of the year, when water can be brought to it. It is owned by private parties, who exact a royalty from the miners, who, when working with rockers, are said to frequently make as high as $10 per day. Rich gravel is also said to have been discovered in some old water courses in the northeast corner of the county, but lack of water has hitherto prevented work being done thereon. There has also been some gold-washing on the western side of the county on the rancho de los Carrisalitos, about twenty miles southwest from Los Banos.

MODOC.

While Modoc may and, no doubt, does contain mineral deposits of many kinds and of much importance, none of ascertained value has yet been discovered. Many years ago a number of silver-bearing lodes were located in the mountains near Surprise valley, and some prospecting work done. On one of the locations a quartz mill was erected, but owing to the remoteness of the place, and, in some measure, to Indian hostilities, the work of development was tardy, and, when the mill was destroyed by fire, it was abandoned. The amount of bullion obtained from the working was inconsiderable, so the extent and value of existing deposits are left, as yet, undetermined. The settlers in the county have turned their attention chiefly to farming and stock-raising; mining is nearly altogether neglected. In Lassen county, just over the southern boundary of Modoc, quartz mines are being worked. Modoc's mineral wealth is yet lying dormant, awaiting the awakening hour of enterprise.

MONO.

It is now about eight years since the business of mining for the precious metals at Bodie, the principal camp in Mono, began to decline. Having taken an unpropitious turn, this industry fell off year by year, until at last nearly every stamp in the district was hung up; exploratory work greatly abated, and bullion production almost wholly extinguished; results due to the exhaustion of the pay ore in the more largely producing mines and the failure to find other deposits of this kind, either in these mines or elsewhere in the district.

For a number of years preceding this break in her fortunes the town of Bodie had been exceedingly prosperous. For this there was a double reason. The output of bullion had been large, while immense sums of money, mostly collected by assessments, had been expended in exploratory work; active prospecting on not less than thirty different claims, all equipped with steam hoisting works, having been kept up throughout this period. The bullion product of these mines during the time they were in bonanza, some six or seven years, amounted to nearly $20,000,000. The yield of the Standard and the Bodie Consolidated from 1877 to 1884, inclusive, amounted to $10,000,000 and $4,000,000 respectively, the bullion of the former consisting of 86 per cent of gold, 14 per cent of silver; of the latter, 68 per cent of gold and 32 per cent of silver. From such large products made in this one locality only eight years ago, the total annual output of the bullion has dwindled to less than half a million dollars for the entire county.

Considerable work, however, is being done in the Blind Spring, Montgomery, White Peak, Indian and other districts. This is all on a small scale, but ore averaging $150 a ton in silver is found. In the Lahe, Homer and Tioza districts are mines upon which work is being done.

here are external gravel deposits which have been successfully worked in the past and which would unquestionably repay systematic development.

MONTEREY.

The existence of gold and silver in Monterey county has been known from the earliest settlement of the country. The principal locality where gold has been found is in the Los Burros mining district in the southern end of the Santa Lucia range, and ledges of auriferous quartz are said to have been found in the Cholame valley, also near Cholore peak in the Gabilan range.

The Los Burros district, which covers a wide area in the southern portion of the Santa Lucia range, was organized in 1876. Prior to 1887 no mineral veins of importance had been discovered, the prospectors principally confining their attention to quicksilver and placer workings. Although several quicksilver claims were discovered, but little was done toward their development. Placer mining was carried on intermittingly for several years. At one time over 100 Chinese were engaged in gold washing in the vicinity of Jolon, it being supposed that the land in that neighborhood was Government territory. It proved, however, to belong to the Milpitas grant, and the owners compelled the Chinamen to discontinue their work. Gold washing was afterward carried on further west in the ravines and gulches of the Santa Lucia range. The gold was principally coarse gold nuggets, some of the value of $5 being occasionally found.

In 1887 W. D. Cruikshank was prospecting in Alder creek, and discovered a "blind lead" from two to four inches wide, containing free gold. He commenced sinking upon it, and the ledge widened out as he went down. He then crushed about twenty pounds in a hand mortar, and washed out $13 worth of gold. He continued sinking, and put up a horse arrastra. This he used for about four months, during which time he realized sufficient to erect a three-stamp mill and a two-horse-power engine. He ran this mill from November, 1887, to the first of June, 1888, when a failure of the water supply, caused by draining the upper workings of the mine, compelled a removal of the mill, and a consequent temporary suspension of milling operations. Further developments of Mr. Cruikshank's mine, which is called the Last Chance, have brought to light five distinct leads, each showing well defined quartz veins, three of which have been found of sufficient importance to work.

NAPA.

While Napa is distinguished as a fruit, grain and vine-growing county, it possesses also a variety of mineral products, of which gold, silver, mercury, iron, chromium and manganese are the principal.

The silver-bearing veins in this county are mostly confined to the lower slopes of Mount St. Helena, where a great many claims of this kind were taken up and much work done upon them from fifteen to twenty years ago. As the developments made proved disappointing, operations gradually ceased, the most of these claims having afterward been virtually abandoned. Although work there has during the past year or two been practically resumed, the only mine that is at present producing bullion is the Palisade, owned by Messrs. Grigsby & Johnson. This property, which comprises four claims, each 1500x600 feet, is located on one of the southern spurs of Mount St. Helena, near the foot of the grade leading from the town of Calistoga into Lake county. A well-built wagon road connects the town with Calistoga, lying about two and a half miles to the south. Two veins have been developed here, one running nearly north and south, the other lying east of this vein, having a trend more to the west.

Gold-bearing quartz veins crop out at many places on Mount St. Helena. Although a good deal of prospecting has been done on these veins, nothing of large value has ever been developed.

NEVADA.

Nevada is one of the imperial mining counties of California, contesting with Amador the honor of being the largest bullion-producing county in the State. The annual output of gold, amounting now to nearly $3,000,000 for each county, would have been much larger but for the suppression of hydraulic mining. The bullion product of Nevada has suffered the largest curtailment from this cause. Every form of gold mining elsewhere pursued is represented in this county, gravel washing by the hydraulic process alone excepted; this, after reaching here its greatest expansion, having been prohibited by the courts. In Nevada county, Cal., gold quartz mining had its origin, the business having begun at Grass Valley as early as 1850, in which year the first quartz mill in the State was erected. In Nevada, also, auriferous gravel washing by the hydraulic method was invented and first practiced, the process having afterward in this county seen its most extensive application. Here are found the longest and most expensive water ditches and the most capacious reservoirs, constructed in this or, perhaps, in any other country. The record made by some of the quartz mines of this county is very remarkable, both as regards large, long-continued and steady production. The ores here are for the most part of good grade and free milling, carrying usually not over 2 per cent of sulphurets. The concentrates yield on an average about $100 per ton. The ore is chiefly gold-bearing quartz, while the veins are not apt to be large, ranging generally from two to three feet in thickness.

In this district the usual vicissitudes have accompanied quartz mining. A number of mines have yielded largely in treasure. There have also been many failures, but they were owing more to the want of capital or unskillfulness in management, as is shown by the fact that mines that failed under one management have yielded profitably under another by the use of improved and better methods.

It may be stated that there has never been a time since quartz mining began in the Grass Valley district, forty years ago, but that one or more quartz mines have been worked at a profit, while a like statement cannot be made for any other mining district on the Pacific coast. From the best obtainable data it is estimated that the quartz mines of Grass Valley have produced over $100,000,000 in gold bullion. This is sufficient to indicate the value and permanence of the quartz lodes of the district and its mining prospects for the future.

Quartz mining has not been so extensively conducted in the Nevada City district as at Grass Valley, but the business has been important there, and is also growing at Willow Valley, an adjoining district, and the future of the industry in that locality is one of abundant promise. The ores of these districts are more heavily mineralized than those of Grass Valley, the ores of the latter yielding more readily to the free-milling process. The concentrates of all the ores of the districts named are generally of high grade.

Taken altogether the quartz mines have produced largely in gold, but it may be said that such mining is as yet only in its infancy here, and there is an inviting field for the intelligent use of capital and labor to enter, with the prospect of abundant recompense.

Nevada county has within its borders an extensive system of ancient rivers. The immense deposits of auriferous gravels of the tertiary cover the greater part of the ridges between the Bear and South Yuba rivers, extending east from Rough and Ready, Grass Valley and Little York to Omega; also on the ridge between the Middle and South Yuba rivers from Smartsville, extending east to Snow Point, is the most extensive and richest deposit of auriferous gravel in the United States, if not in the world. These immense auriferous deposits are covered in places with volcanic capping.

Prior to the anti-debris litigation thousands of men were employed working the auriferous deposits by hydraulic process. Millions of dollars of capital were also invested in the construction of canals, pipe lines, and long sluice tunnels for working the deposits by hydraulic process.

The auriferous gravels in Nevada county, if worked by hydraulic process with all the water available, would, at the lowest calculation, yield $5,000,000 or $6,000,000 per year and could not be exhausted in a century.

The hydraulic mines throughout the county are all closed by injunction; the majority of the water ditches and canals are going to ruin, and the little towns and villages dependent on the mining industry are all about deserted and going to ruin.

Since the cessation of hydraulic mining, several attempts have been made to work the bottom gravel by drifting, but in most cases it was found to be of too low grade to pay for the handling by the drifting process.

ORANGE.

Orange county is well supplied with valuable minerals. There are some apparently extensive silver deposits in the Santa Ana range, and both gold and silver in some other portions. What is known as the Pilligrin or Alma "diggings" are on one of the branches of the Santiago creek; they crop out on one side of the mountain, the upper portion composed of surface pockets and chimneys, with indications of a fissure vein below. The elevation here is about 2300 feet. Several tunnels have been excavated and much good ore has been extracted. On the opposite side of Santiago canyon is an elevation called "Carbonate hill," which seems to contain much valuable mineral. It is approached from the southwest along Weakly canyon, and has an elevation of 2600 feet above sea level. The most valuable mineral of this "hill" is lead carbonate. W. S. Morrow, who has taken up several claims, has made openings which expose the ledge for some 3000 feet, and it is said to run *high in silver. The hanging wall is quartzite, and the foot wall is granite. The dip is eastwardly, which is true of all the gold and silver bearing rocks of the Santa Ana range.

PLACER.

Placer has from the first been noted for the varied character and the extent of her mining operations and her large bullion production, the latter having at one time amounted to several million dollars per annum. Of late years the output of gold has been greatly diminished through the stoppage of hydraulic mining, formerly prosecuted here on a large scale. Meanwhile, however, drift gravel mining has been somewhat increased. This branch of the business is now largely carried on in the county, the Forest Hill divide being the site of its most extensive operations.

Mining for gold has been the leading industry and source of wealth of the c unty in the past since the discovery of gold in 1849, and will certainly continue to be one of the principal industries and sources of wealth in the future. The shallow placers were extensive, extending from the lower plains almost to the summit of the Sierra, and were among the richest in the State.

Quartz mining has also been carried on to a greater or less extent since the erection of the Crœsus mill—one of the first stamp mills in the State—on the Crœsus mine in Baltimore ravine, near Auburn, in 1851.

As the shallow placers and river bars were exhausted miners turned their attention to drift and hydraulic mining in the deep auriferous gravels of the ancient river channels on the hills; and whenever the beds of the ancient rivers were accessible for working by shafts and tunnels they were worked by the drifting process and yielded large profits.

Hydraulic mining began in 1854 and was carried on successfully at Yankee Jim's, Forest Hill, Bath, Michigan Bluff, Iowa Hill, Wisconsin Hill, Gold Run, Dutch Flat and other places throughout the county, and as an industry increased in importance and flourished until the debris litigation resulted in stopping, by injunction, all the hydraulic mines in the county. Since hydraulic mining ceased, in 1886, miners and capitalists have turned their attention to the development and working of quartz mines and the opening of the deep, lava-capped auriferous gravel channels for mining by the drifting process. This work is not objectionable, as it can be conducted without doing damage either to the navigable streams or valley lands.

Taking into consideration the fact that there are within the county limits about 200 miles of unworked auriferous gravel channels and an immense area of auriferous metamorphic rock, in which are great numbers of veins of auriferous quartz, and basing an estimate of the amount of gold yet remaining in the unworked channels on the results obtained from channel workings at Forest Hill, Iowa Hill, Deadwood, Last Chance, Canada Hill, Dutch Flat and some deep channel workings between Rocklin and the American river, varying from about $100 to $1000 per linear foot of channel worked and equal to a product varying from about $500,000 to $5,000,000 per mile, it is evident that the amount of gold already extracted is but a trifle compared with the amount remaining in the ancient river channels and quartz lodes.

PLUMAS.

From an early period placer mining has been largely engaged in in Plumas, though this branch of the business has suffered marked curtailment through the cessation of hydraulic operations. Drift and quartz mining, however, continue to be actively and successfully pursued. The principal centers of quartz mining are the Greenville, Dixie and Jamison Creek districts, in the vicinity of Indian, Mohawk and Genesee valleys. Drift and other placer operations are mostly confined to the central and southwestern parts of the county. Deposits of coal, copper-bearing veins and beds of marble rank also among the mineral resources of Plumas.

For a number of years this county held a position in the front ranks of the bullion-producing sections of California, and while she still contributes her quota to the yearly amount, it is on a greatly diminished scale. Two causes have combined to bring about this condition of affairs; on the one hand the injunction placed on hydraulic mining, on the other the mixing up of stock gambling with the management of quartz mines, which has been quite prevalent in some parts of the county. It is stated that the loss to Plumas county through depreciation of mining property and diminished gold production amounted to about $400,000 per annum, and as Plumas county has a population of about 7090 inhabitants that means a loss of over $500 per capita for the entire population.

Among the prominent quartz mines of this county are the Plumas, Eureka, Lucky S., Crescent, Green Mountain, Altouna, Cahalan, Pennsylvania, Indian Valley, Genesee Valley and many others. There are few portions of the State that present better opportunities for capital than Plumas county.

SACRAMENTO.

Some of the richest placer mines that were worked in the early days were located within the boundaries of this county, notably at Mormon Island, Michigan bar and other places. Where the American and Cosumnes rivers debouch into the Sacramento valley from the foothills, placer mining was carried on in an early day. It is still carried on to some extent in the foothills of Sacramento county, where water can be obtained. The extent of placer mining in the neighborhood of Folsom was the subject of inquiry by the Board of Supervisors in the summer of 1890, and their investigations extended over six or eight miles along Willow creek and Alder creek, and the land belonging to the Natoma Company in Granite and Natoma townships. They found about sixty or seventy men engaged in placer mining and drawing their water supply from the Natoma Water and Mining Company, and realizing about $36,000 per month, judging from the amount of gold sold in Folsom. Messrs. Finch & Co., Salvador & Co., P. Carroll and some Chinese were working on land belonging to the Natoma Company that was worked in the fifties by drifting. It consists of alluvial soil and pebbles for about fifty feet in depth, underlaid by cement and a clay stratum. Messrs. Finch & Co. were paying $15 a day for water and appeared to be getting fair returns.

Two miles south from Folsom some drift mining is being done at Rebel Hill, on property belonging to the Natoma Water and Mining Company; eight or ten claims are being worked through shafts forty-five to fifty feet deep. The pay gravel lies upon a stratum of conglomerate, similar to that at Folsom. The pay streak varies from three to eight

THE FORTY-STAMP QUARTZ MILL.

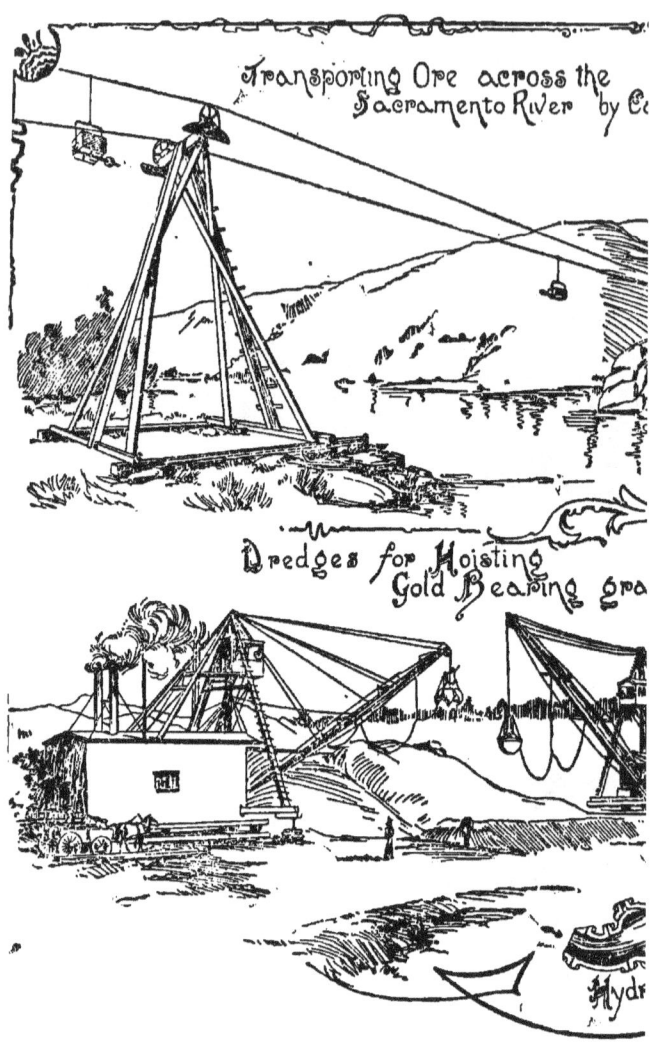

VARIOUS DEVI

feet in thickness. The gravel is hauled to water and washed through sluices, rockers and toms. The owners of the ground own also the water.

SAN BENITO.

This county has never been known as a mineral section, yet there are large deposits of various kinds, some of which have been profitably worked. Auriferous placer workings have from time to time been found and good float rock discovered in the hills to the north and northwest of Panoche, but the locality is too far from water to be available for placer mining during the greater portion of the year. Several years ago a party of Frenchmen commenced packing the dirt in sacks on their shoulders to the nearest creek, a distance of about two miles, but they could only make from 50 to 75 cents per day, and abandoned the enterprise. It is also claimed that a ledge of rock has been struck in the same district that assays high both in gold and silver.

The McLeod district is the principal mineral producing section of the county.

SAN BERNARDINO.

Mining in one form or another has been pursued in this county from an early day. At first the business consisted of placer operations, carried on in its southwestern part, chiefly in Bear and Holcomb valleys, and along some of the creeks and gulches still further west. Later on some quartz mining was undertaken in this section of the county; which were not, however, attended with much success. After the discovery of the Comstock lode, the attention of the mining public having been strongly directed to silver, the whole northern part of San Bernardino was explored for that metal, and with such encouraging prospects that a number of mining districts were organized and many lodes taken up in that region of country. The districts so formed consisted of the State Range, Washington, Argus. Telescope, Armagosa, Potosi and the El Dorado—the Ivanpah district, in the same region, having been formed in 1868, a little later. Owing to their remoteness, the cost of transportation, and otherwise adverse conditions, mining made so little headway in these districts that they came, in the course of a few years, to be about deserted. Since the construction of a railroad across the Mojave desert, rendering the mines more accessible and greatly reducing the cost of transportation, mining operations have undergone some revival in these districts, with the prospect of becoming still more active in the early future.

Meantime the Calico country, much more accessible and every way more advantageously situated, having been discovered and opened up, promises to plant in the very center of this county a prosperous and permanent silver mining industry.

This camp is situated about seven miles north of the town of Daggett, on the Atlantic and Pacific railroad. Notwithstanding the want of fuel and timber and the scarcity of water, the ease with which the ores here can be mined and milled insures a reduction of the cost of working in the near future, or whenever large and well-equipped mills shall be erected, freight and wages reduced and adjacent mines consolidated. At the present time ore is milled at a cost not exceeding $4 per ton. A considerable number of "chloriders" find profitable employment in the numerous mines within a radius of five miles from the town. These cannot afford to handle ore carrying much less than forty ounces to the ton, having usually to pay a tribute of one-fifth, in addition to the expense of sacking, freighting and milling; as a consequence large quantities of comparatively high grade ore are left in the mines or on the dumps, to be worked at some future time. The depression in the silver market has exercised a bad effect upon the Calico mines and many good properties are now closed down.

San Bernardino contains the following mining districts, nearly all located in the desert portion of the county: The Gold and Silver, Morongo, Brier, Holcomb Valley, Borax Lake, Ruby Mountain, Twenty-nine Palms, Ibex, Borrows, Morongo, Solo, New York, Exchequer, Grapevine, Temescal, Ord, Black Hawk, Trojan, Silver Mountain, Lava Bed, Alvord, Calico, Clark and Scanlon.

The copper, lead, zinc, asbestos, iron and other minerals are also found in this county.

On Lytle creek, thirteen miles northwesterly from Colton, auriferous gravel is found. From the mouth of the canyon northerly, five miles, to Pratts, there is more or less of it, and considerable work was once done here. The available places, however, are nearly exhausted and work has been discontinued for a long time. At Texas Point some $80,000 is reported to have been taken out by hydraulic process. Above this point the ravine spreads out in flats, covered by large granite bowlders.

Crossing the ravine near Glenn's ranch, township 2 north, range 6 west, section 22, a vein of gold-bearing rock occurs, which gave a good horn-spoon prospect, showing free gold; but no work has been done on it other than sinking a five-foot shaft.

On the headwaters of the San Gabriel river, close to the line on the Los Angeles side, and westerly from the point above mentioned, ten miles from its source southerly, and thirty miles westerly from Glenn's ranch, on the western slope of the San Gabriel and Lytle creek ranges, lies the San Gabriel gold mine. A shaft twenty feet in depth has been sunk, exposing a vein twelve feet wide, and a quartz mill is being built on the property.

On the northern slope of the San Antonio Peak, at an elevation of 8140 feet, a dead river channel of auriferous gravel was discovered in the summer of 1882. In many respects this gravel resembles that of the pliocene beds so extensively worked in the middle and northern counties of California.

SAN DIEGO.

San Diego county, besides a variety of other useful minerals and metals, possesses a considerable wealth of gold and silver, chiefly the former, her auriferous resources consisting of both vein and placer deposits, the latter not extensive. Salt is also produced in this county. Gold-bearing quartz lodes were discovered here as early as 1869, the site of these first discoveries being in the Cayumaca mountains, a high range distant some sixty miles from the coast. The Julian, the Banner and several other districts were afterwards organized here, many claims taken up and much work done, this still continuing to be the principal quartz locality of the county.

From statistics prepared by Chester Gunn of San Diego it appears that the gold yield from these districts, from their discovery in 1869 up to 1880, was over $2,500,000. The surface rock being rich, it was a good camp for poor men. That this neighborhood has not yet been properly explored is shown by the fact that in the early part of 1890, in a little valley five miles southeasterly from the town of Julian (at an altitude of 4700 feet), seven locations had been made on small, rich veins of auriferous quartz, the croppings carrying free gold to such an extent that it seems almost impossible that for nearly twenty years it should have escaped the eye of the prospector.

The leading mine of San Diego county is undoubtedly the Stonewall, which is situated in Julian district. The yield of this mine has been very large, and it is still believed to be a good property.

Speaking generally of the history and prospects of gold mining in San Diego county it may be said that while a good deal of prospecting and surface scratching has been done at various localities, yet the total aggregate amount of intelligent and systematic mining which has ever yet been done within the limits of the county is extremely small.

This has been due to a variety of causes: First, most of the mines are situated at considerable distances from any points which have hitherto been very easily accessible to travelers, and very little has been known about them outside of the county itself. Second, there has existed from the beginning a widespread but unreasonable and unfounded prejudice against the county, which has rendered it almost impossible to induce capitalists to invest any money in mines that are located there. Thus, most of the mines have in the past been owned and worked by men who were comparatively poor and had not the requisite means to properly develop them, which accordingly they failed to do. Other mines have shut down for other causes, which were not the fault of the mines themselves, such as unskillful and incompetent management (which will ruin any mine); costly litigation, which always arises to a greater or less extent wherever rich mines are found, etc. Yet many of these mines have yielded large sums in the past, and some of them are to-day running and doing well.

SAN FRANCISCO.

The county of San Francisco is to be numbered among the gold-bearing localities. Located in the western part of the county is a gold-producing beach, which, commencing at the outlet of Laguna de la Merced, extends thence south along the seashore for a distance of about two miles. Nearly all the gold here occurs in strata of magnetic iron ore, the so-called black sand, there being very little in the ordinary sand of which the beach is mainly composed. The gold found consists of minute particles, much of it being of almost atomic fineness. A piece as large as the head of a pin has never probably been washed out here.

Like all auriferous beaches, and most other placers, this is a secondary deposit, the original sources of this gold having been the quartz lodes that formerly existed in the basin that has its drainage into the laguna. Some have assigned for this gold another and more distant origin, advancing the theory that it was by ocean currents brought down from the north and here thrown up and left by the surf. This, however, was before the country adjacent had been examined and its auriferous character established.

Gold-bearing quartz veins and their attendant metamorphic rocks are found not only in the basin of the Laguna Merced, but throughout the entire San Francisco peninsula, and even along the Santa Cruz branch of the Coast range, all the way down to the bay of Monterey. A quartz mill was put up in these mountains many years ago, and for a time run with some success. A nugget of gold weighing several ounces was picked up in that vicinity at an early day. Careful prospecting along all the ravines and arroyos throughout this region reveals frequently a speck of free gold, with many grains of the characteristic black sand.

SAN JOAQUIN.

While San Joaquin is not and never has been a mining county, yet it does not follow that there is no gold there. On the contrary it is certain that the precious metal does exist in the gravel of the streams which pass through the county, all of which have their sources in the richest placer mining region of the early days. Some little mining has been done on these streams within the limits of the

—g nty, and it may well be that in the future this may become a source of considerable wealth.

SAN LUIS OBISPO.

Gold, silver, lead, copper, quicksilver, chromite, gypsum, onyx, salt, lime, coal and petroleum have been found in the mountains of this county. It is a matter of history that gold was shipped from San Luis Obispo and neighboring counties prior to its discovery by Marshall in 1848. The explorers of the Pacific railroad reported gold west of Salinas in 1854, though its existence in the San Jose mountains had long been known. Gold has been and is still washed from sands in the bed of the San Marcos creek, about four miles northwest of Paso Robles, during the wet months of the year, yielding, it is said, as high as from $3 to $4 a man a day. Placer claims have also been worked thirty miles southeast of Templeton since 1870-71, ground sluicing and panning, when water has been plentiful, having yielded from $2 to $4 a day.

The placer mines of the La Panza district are the best known and are probably of the most importance. They are situated at the southeastern foot of the San Jose range, which rises as a formidable mountain joining the Santa Lucia, and over $100,000 in gold has been taken out. During 1878 there was quite a rush to these parts and prospecting was carried on in nearly all the gulches leading from the San Jose range to the San Juan river. The chief interest was centered in the De la Guerra gulch, where the most mining was done, even as late as 1882; also upon the Navajo creek, which is a stream of constantly flowing water. Some of these placers have yielded as high as $4 per day. The gold was coarse, pieces worth 50 cents or 80 cents being of frequent occurrence. Haystack canyon also has running water and gold. Near the head of this canyon are falls of twenty feet, where the water descends into a basin twenty feet across and ten or twelve feet deep.

These streams reach the channel of the San Juan during very wet weather. Of late years these mines have not been actively worked, chiefly on account of the scarcity of water. In the southern portion of the county gold has also been found in sands on the seashore in considerable quantity. They are reported as yielding from $1 50 to $2 per day to the miner, and, as the gold dust appears to be renewed by the washing of the sea, the deposits are practically inexhaustible.

SAN MATEO.

The minerals of this county, as far as investigation is shown, consists of gold, silver, petroleum, coal quicksilver, lime and building stones. Of these petroleum and building stones are at present alone turned to any practical account.

Traces of gold have been discovered in various creeks and gulches in San Mateo county, especially on the Hawes ranch, near Redwood City, prospects there having been struck which yielded several colors of gold to the pan. There is said to be a quartz ledge on Deniston peak which assays a few dollars to the ton, and from which specimens showing free gold have been obtained. Also upon the ranch of Ole Durham, on the Tanitas creek, is a ledge of quartz which is said to assay well. Placer mining has at intervals been carried on at several points along the seashore with varied success. A bed of black sand on the beach at the Deniston ranch, about one mile north of Amesport landing, was worked with only partial success, though one of the parties stated that he recovered about $7 to the ton.

SANTA BARBARA.

Gold, silver, quicksilver and other valuable minerals are found in this county. On the San Marcos ranch there is said to be a lode that assays well in both gold and silver. Gold-bearing rock has also been found on the Buel ranch near Los Alamos. Placer claims have been worked at Pine mountain, also at the headwaters of Zaca creek, and at several places in the San Rafael mountains. A few colors of gold are occasionally found in the creeks running from the Santa Ynez range. Gold-washing has also been carried on upon the seashore. The most successful operations were at Point Sal, in the northwestern corner of the county. Gold-washing has been intermittingly carried on here by the Point Sal Mining Company. The gold is found in streaks of black sand from three to four feet below the surface of the beach. They run from one inch to two feet in thickness, usually being about one foot, and from thirty to forty feet in length. The bank of the beach runs north and south, the streaks of sand east and west toward the ocean. Beneath the black sand is blue clay in some places and sandstone in others. The richest deposits are found on the sandstone where it is worn into ridges, being favorable to the concentration of the gold. This sand is run into a hopper, where a stream of water carries it over amalgamated plates. About twenty-five tons of this sand yielded $137.

Seven miles north of Point Pedernales and twelve miles west of the village of Lompoc is a long beach where gold in considerable quantities has been obtained from the washing of the sands. The auriferous ground extends northerly some two or three miles to the opening of the valley of the Santa Ynez river, two miles south from the mouth of the stream. Through this extent the bluff is from twenty to thirty feet high, being a cemented mass of sand and gravel in horizontal layers as if it were the channel of an ancient river. At the base of this the sand beach of the ocean slopes away to the water in a width of from 100 to 200 yards. The waves at times beat against the bluff. During heavy storms the light gray sands are washed away, leaving a surface of black ferruginous sand, which is accompanied by fine particles of gold and platinum. In other storms the gray sand is returned to a depth of four to six feet over the black sand, and this is the usual condition.

SANTA CLARA.

This county ranks well with any other in California in the possession of valuable mineral resources. Prospects of gold and silver have been discovered in the creeks of the Mount Hamilton group and Santa Cruz mountains. Gold was discovered in the bed of the San Francisquito creek, near Mayfield, eighteen years ago and a placer camp was started, but was soon abandoned. Gold was also panned out in the pioneer days in Coyote creek, within the present limits of the city of San Jose, but in trifling quantities. Specimens of silver-bearing quartz, purported to have been discovered in the vicinity of Mount Hamilton, have also been brought into San Jose. Mr. Hahn, who lives in Alameda township, states that he discovered a quartz ledge in the Coast Range; that the ledge runs northeast by southwest, and that the croppings, which are three feet wide, assayed $4 in silver and 80 cents in gold per ton.

SANTA CRUZ.

The metals and minerals as yet discovered in the county are gold, silver, coal, bituminous rock, quicksilver and lime, together with sufficient building stone for local purposes. Of these gold and silver, petroleum, in the form of bituminous rock, lime and building stones are all that can be counted as known sources of actual mineral wealth.

Placer mining is carried on along various creeks in this county when water is abundant and generally yields fair wages to those engaged in it. The sluice and rocker have been familiar sights both on Wardell creek and at Gold Gulch, near Felton, and on Major creek, on the ranch of J. L. Thurber. At the latter place three men within thirteen days took out thirteen ounces of gold with sluices and plain riffles, no quicksilver being used. The gold was coarse and rough, some pieces being attached to rose-colored crystalline quartz. The largest pieces ranged in value from 25 cents to $10, the majority being the size of a grain of wheat; there was no fine gold. The gold was worth $18 per ounce. Parties have mined on the creek for five years, and it is stated that over $28,000 worth of gold has been shipped through Wells, Fargo & Co. at Santa Cruz. In early days a bowlder, the size of which was estimated at some sixteen cubic feet, was discovered in Gold Gulch, near Felton, which, when milled, yielded some $38,000. A small mill was erected on the spot, but, although a tunnel 1200 feet long was run into the hill in hope of striking the ledge from which the bowlder came, nothing remunerative was developed. Various attempts have been made to discover the source of the placer gold, but heretofore prospecting has met with but little success in the county, the great obstacle being the depth of soil covering the rock formation and the dense growth it maintains, both of which prove a hindrance to the prospector and geological observer. yet some few ledges have been unearthed. Considerable attention has been paid to the auriferous black sand which occurs in the ancient raised beaches of Santa Cruz, upon which abortive experiments have heretofore been made; also the auriferous sands upon the seashore, which in slack times have yielded small wages for manual labor since the early settlement of California.

SHASTA.

This is one of the counties that ranked among the largest gold-producers in the early history of placer mining, and it had many extensive deposits of gold and silver ore, as well as other minerals. Some little placer mining is still occasionally done, but the bulk of mining now carried on is the development of the quartz deposits. Old Diggings district has several notable mines, including the Texas and Georgia, which has produced ore that paid $240 to $290 to the ton in gold and $10 in silver. The Utah and California averages $150 to the ton, and a large amount of work has been done on it.

The Lower Springs district, a few miles northwest of Redding, has half a dozen good veins which show rich ore. In early days the ravines and gulches in the mountain through which this vein crosses were exceedingly rich in "placer," mainly below the point of outcrop of this ledge. During the past year several pieces or nuggets were picked up, having been washed down by the winter's rain. The owners of the properties will not permit any mining, as the level lands adjoining are under cultivation—vineyard and orchard.

One of the best known mines in Shasta county is the Washington, located at French Gulch. This mine was located thirty-nine years ago. The first stamp mill erected in the county, containing six stamps, was built on this claim. The mine has produced between $500,000 and $600,000 since, but is not at present a paying proposition. The owners are prospecting the property for the purpose of opening out new ore bodies. It has an elevation of 2000 feet above sea level and is two and one-half miles west from the town of French Gulch.

The Niagara mine, at French Gulch, is another notable property which has been worked for a long time and has paid largely.

In nearly all those portions of the county

where placer mining was ever followed the quartz miner is at work, and many promising ledges have been discovered. These ledges are not confined to any one part of Shasta, but are met with over a large part of the county. One of the most noteworthy deposits of mineral in the county, and in fact in the world, is the Iron Mountain, fourteen miles northwest of Redding. There is a good road to this mine leading through Shasta and then into the mountains, over a narrow and steep grade, and a visit to the remarkable deposit is well worth undertaking. The Iron Mountain mine has a twenty-stamp mill constantly at work, and large shipments of silver are regularly made. There is nothing peculiar about the mill, the ore being roasted and worked by what is known as the pan process. The mine itself, however, is a curiosity. There is a solid mountain of ore rising some 1200 feet above the gulch in which the mill is located, and extending for miles in either direction. No shafts or tunnels are needed in working the mine, but the ore is simply quarried from the face of the mountain and sent to the mill through chutes. When first discovered the ore was supposed to be a deposit of iron, and so indeed is about 75 per cent of it. But assays showed that there was a large proportion of silver, together with a little gold and copper. Consequently it was decided to work the ore for the precious metal and let the rest go, and this is now being done.

Underneath the iron and silver combination, however, are immense deposits of sulphurets, which run from $30 to $150 in silver, and which are largely of such a nature that the ore can be shoveled out like so much loose sand. Prospect tunnels have been run in these sulphuret deposits, and no limit to their extent has been found. The mine, had it been discovered on the Comstock or in any other noted mining region, would be one of the wonders of the world and would produce a second Washoe excitement. But it is the property of a few men who are contented with working along in a quiet way, and, while there are unquestionably millions in it, it is not for sale, and the stock gamblers have had no hand in it. The capacity of the mill is to be doubled shortly, and indeed, there is ore enough in sight to keep the largest mill that could be put up profitably running for an indefinite period.

Another mine which has attracted considerable attention is the tellurium deposit discovered a few years ago some three miles from Redding in a gulch that is tributary to the Sacramento river. This mine was discovered by Peter Scherer, who was led to search for a quartz ledge because of the rich placers that had been found in the vicinity. The first shot put about the year 1850 in the San Joaquin river and its tributaries, and a mining population of about 1500 sprung up in that vicinity. Many men made small fortunes, while the degree of success of others was less. One of the claims there paid its proprietor $117,000. There were also several Chinese companies engaged in mining who faied very well. At Fine Gold creek alone there were at one time in the early fifties 500 people mining. The placers were worked for all they were worth until the winter of 1807-68, when the heavy floods swept everything away, discouraging the miners, the greater number of whom sought more promising and newer fields. The advent of the agriculturist about this time and the exhaustion of the old placers also contributed to bring about the cessation of placer mining. There are still some very good claims there, but they have never been worked on account of the difficulty of turning the water. Some quartz mining was done early in the fifties. The quartz mining belt is from eight to ten miles in width, running parallel with the Sierra Nevada, its edge being about twenty-five miles from Fresno. The various mining districts are Hildreth, Auberry, Mount Raymond, Fresno Flats, Grub Gulch, Coarse Gold, Fine Gold, North Fork and the Minarets. Many fine prospects have been made, which will no doubt prove valuable properties when developed. When experienced mining men become interested and invest capital—the scarcity of which up to the present has prevented the full development of the mines—the results will be most gratifying. The ores extracted have in many cases assayed very highly, as much as $500 and in some cases even $1000. Of course, this is much above the average, but the ore is of a high grade generally. The capital invested in the mines in this county amounts to fully $350,000, while the number of men employed is about 500. Though many of the mines have been worked quite steadily, the operations, except in a few instances, have not been on a very extensive scale.

HUMBOLDT.

Humboldt county possesses some valuable gold mines, notable being those at what is known as Gold Bluff, where there are extensive black sand deposits. These deposits created much excitement some thirty-five or thirty-six years ago. They extend from Oregon eight miles in a southerly direction. The whole line of the beach was worked for the gold the sand contained. This deposit of black sand contains traces of platinum, osmium —iridium too minute to have any commercial value, but with sufficient gold to make the business of gathering it lucrative. Several parties engaged in this auriferous harvest have retired with a competency. From the date of discovery to the present writing these beaches have been annually worked, success depending, time occupy a prominent place among the gold-bearing sections of the State.

SISKIYOU.

All kinds of gold mining known in California are successfully prosecuted in Siskiyou county—quartz, placer, drift and river. A great deal of hydraulicking is also done, as the inhibition against that industry does not apply here. On the Klamath river, as also on its main tributaries—the Shasta river, Scott river and Salmon river—the main bullion-producing sections of the county are found. The greater proportion of the mineral wealth is obtained at present from the gravel benches, bars and ancient river channels, not that the county is lacking in vein deposits, but her mountains are so rugged and precipitous that transportation of large machinery as required to develop quartz mines is difficult. As the debris from the hydraulic mines is dumped into the canyons and streams tributary to the Klamath river, which is torrential and discharges directly into the sea without harming any large spaces of arable land along its course, no damage is done to the farmers, and no objection is made to this method of working.

The mining districts in this county are the Cottonwood, Yreka, Humbug, Deadwood, Oro Fino, Callhan's Ranch, Scott River, Oak Bar, Sciad Valley, Cottage Grove, Liberty, South Fork of Salmon and Forks of the Salmon.

There are a number of good quartz mines in the county which have paid well and there is any amount of opportunity for the investment of capital.

The following claims are in operation whenever conditions are fitting, particularly the supply of water, on which this interest is entirely dependent; Hydraulic claims, forty-seven; wing dams, twenty-two; drift claims, twenty-three; sluicing claims, seven; employing 942 operatives. This is an underestimate of the actual number of men employed in the mines of Siskiyou county. A large number of men are engaged in prospecting and opening claims who have not been taken into account.

The quartz interest of the county is yet in an embryo state. Comparatively little attention is paid to it, but it is beginning to assert itself, however. There now sixteen mills in operation, dropping 120 stamps. As in the case of gravel, the output of gold from the various mills is extremely difficult to arrive at. The number of men employed is about one hundred and twenty—or one man to a stamp. This would swell the number directly engaged in mining in this county to 1062.

SONOMA.

Although Sonoma is among the foremost vine and grain-growing counties of the State, its mineral wealth is by no means inconsiderable, both the royal and several of the more common metals being counted among this class of its resources.

The gold deposits here consist mainly of auriferous sands on the ocean beach, and some placers of limited extent along the confluents of Russian river. In excavating near Tyrone, on the line of the North Pacific Railroad, some small stringers of silver-bearing ore, associated with magnesian shales, have been cut. Argentiferous indications are reported elsewhere in the county. One and a half miles east of the same locality is situated the Sonoma copper mine.

A little gold is also said to have been found almost everywhere in the gulches among the hills for a considerable distance both north and south of Cloverdale, on both sides of the Russian river valley; and it is said that at a few localities considerable placer mining was once done, although it never paid much more than ordinary wages.

STANISLAUS.

Stanislaus county cannot at the present day boast of any large store of mineral wealth. Formerly a good deal of gold dust was taken from the bars along the Stanislaus and Tuolumne rivers, in the northeastern part of the county, but these bars have become so much depleted that they afford now employment to only a small number of men, and there being in the county, so far as known, no other deposits of the precious metals, the output of bullion has of late years amounted to comparatively little.

Some placer mining is still being carried on upon the Stanislaus river, near Knight's Ferry, in the alluvial deposits which skirt Table mountain upon the southwest, principally by Chinese. Several parties are also working on Goat island during such times as they can obtain water from Little John creek. On the Tuolumne some work is being done by the La Grange Ditch and Hydraulic Mining Company. This company, about twenty years ago, purchased the title to all the gold-bearing lands between La Grange and Patricksville, including French Hill.

SUTTER.

In the early history of this county gold was found in the ravines of the Marysville or Sutter buttes, and several thousand dollars' worth of dust was washed out. The ledge from which this dust came does not appear to have been discovered.

TEHAMA.

Little mining is being done in this county at present. In former days some river mining was carried on in the upper reaches of the Sacramento river, which runs through the county, but that has pretty well ceased, and the only kind done to speak of is near the western boundary of the county, in the Coast Range mountains, where some large deposits of chrome iron have been developed. Gold has been found in some of the streams that have their rise in the

Sierra Nevada mountains, but it is many years since enough mining has been done to amount to anything.

TRINITY.

From the very earliest history of California Trinity county has been known to be rich in deposits of gold, both placer and quartz. As early as 1845 Major Redding visited this section, and he must have discovered evidences of the existence of gold, for when Marshall made his discovery at Coloma in 1848, Redding at once set out for the Trinity region. Crossing the mountains at the head of Cottonwood creek, he came upon the Trinity river at a point where the creek now named Redding empties into the Trinity. Quoting from the Major:

I prospected for two days and ound he bars rich in gold; returned to my home on Cottonwood, and in ten days fitted out an expedition for mining purposes; crossed the mountain where the travel passed two years since from Shasta to Weaver. My party consisted of three white men, one Delaware, one Walla Walla, one Chinook, and about sixty Indians from the Sacramento valley. With this force I worked the bar bearing my name. I had with me 120 head of cattle, with an abundant supply of other provisions. After six weeks' work parties came in from Oregon, who at once protested against my Indian labor. I then left the stream and returned to my home, where I have since remained in the enjoyment of the tranquil life of a farmer.

Following the discovery of Major Redding came prospectors from all sections for gold, working the river bars, the ravines and gulches, extracting the gold from the gravel and sands by the rocker, tom and sluice. The evidence of these early workings can be seen along the course of almost every streamlet, creek, gulch and ravine tributary to the Trinity.

The wealth of Trinity county is in its gravels, the ancient channel and the high benches of present waterways. Quartz veins carrying gold are being prospected and worked in different sections. Others that have been opened and worked for several years have yielded and are yielding handsome returns to the owners.

There are seventy-four hydraulic mines in this county, and whenever there is sufficient water a large amount of gold is extracted.

TULARE.

Of the streams that drain the western slope of the mountains in Tulare county only two of any size have failed to yield placer gold. These streams are the Kaweah and the Tule rivers. North of the White river there is scarcely any evidence of early prospecting to be met with. But on the head waters of the middle fork of the Kaweah is Mineral King district, sixty miles northeast of Visalia by road, the discovery of which created a great excitement nearly twenty years ago. There was no placer gold reported, but there were many mineral-bearing veins claimed to be rich in gold, silver, lead and zinc in veins of limestone. About 1875-76 efforts were made to work these mines, but soon abandoned.

On Rattlesnake Peak is a slate formation in which is embedded great quantities of pebbles of mica slate, hornblende slate, quartz and granite; and when this rock is decomposed, placer gold is found in the gulches, showing it to be among the oldest of the gravel deposits.

High in the Sierra near Mount Bruner, possibly in Inyo county, are a number of veins owned by Messrs. D. Ilidei and Soto, which produce very rich specimens of ore bearing gold, silver and copper. These have been partly opened and some excellent ore taken out, but the inaccessibility of the region has prevented their development.

On White river D. W. Grover of Santa Cruz owns the Mammoth mine, on which he has erected a five-stamp mill. Specimens of the ore show free gold. The result of the working has not been told.

Messrs. H. B., E. B. and O. D. Barton and J. S. Butts have located a long series of claims of gold-bearing rock in the vicinity of Rattlesnake creek.

TUOLUMNE.

Tuolumne is one of the leading mining counties, and doubtless always will be. It is crossed by the mother lode in the western part, and numerous rich quartz deposits have been discovered over a wide area, while there are many gravel channels that will repay working. Some of the richest placer mines ever worked in the early times were in this county. At present quartz mining is about the only branch carried on to any extent, though something is still done in the primitive working of placer deposits. This county, in the immediate vicinity of Sonora—that is to say, within a radius of several miles —is noted for the great number of "pockets" of gold that have been taken out. Bald Mountain, and in the vicinity of the Bonanza mine, have produced a greater number of pockets, varying in value from a few hundred dollars to many thousands, than any other mining section in the world. "Jackass Hill," about four miles northwest in an air line from Sonora, is also a noted pocket district. The chief of all the noted veins of this character is undoubtedly the Bonanza. In the neighborhood of $2,000,000 have been taken from this mine, and the judgment of experienced miners in this branch of mining is that all "signs" indicate further successes in it in the near future. The fissure is twelve feet in width, and contains three veins of quartz; the foot wall vein is about four inches in width, the hanging wall vein about the same width, and the middle vein averages thirteen inches.

There are a great many pocket mines being worked on Bald mountain, the principal ones being the Ford, the Austrian, the Wilson, the Garrett, and the Sugarman. All have had varying results. Quartz claims are being worked in many portions of Tuolumne county and the opportunities in this line are extensive.

VENTURA.

Besides other minerals Ventura county contains deposits of gold which are of no small importance. Most of these are found in the Peru district, which is several miles in extent, the most important portion lying in Ventura county. Gold was discovered here long before the gold excitement of 1849. Professor Whitney says it was somewhere in this vicinity that gold was first obtained in California in considerable quantity, and that was as early as 1841. M. Duflot de Mofras says that the locality was in the mountains six leagues from San Fernando and fifteen leagues from Los Angeles where gold was first discovered. Bancroft makes mention of the fact of this locality having been worked more or less during the first half of the present century. It is evident that the yield of gold and silver at this locality has amounted to a large sum in the aggregate.

The principal lode in the Piru district is called the Fraser mine. During the time it was worked, a period of eight years, until October 31, 1879, because of litigation arising from disputed ownership, it is believed to have yielded about $1,000,000 in gold. The difficulty is now said to be on the eve of settlement, and it will be worked by improved methods and on a larger scale than heretofore.

YOLO.

Placer mining has been carried on in a small way along the foot of the Coast Range in Yolo county, and quartz, that by assay showed a small amount of gold and silver, is said to have been discovered farther back in the mountains. In an early day a mining camp for some time maintained a struggling existence near the mouth of Putah creek. Some sluicing is also occasionally done in the foothills to the west of the Orleans vineyard, near Capay, during the winter when water is plentiful, and it is said that as much as $2 per day have been made to the man.

YUBA.

With an area of 625 square miles this county in its eastern part contains a few of the largest and most extensively developed hydraulic mines in the State. Some of these have, in the last few years, been developing into drift mines. The most notable are along the southern bank of the Yuba river in the neighborhood of Smartsville, where a part of the blue lead was traced as far as Timbuctoo; there it appeared to have been raised up and cut off, at least it was never found beyond that point. Towards the northeastern part of the county in the foothills some good quartz veins are being developed, and the one time extensive mining camp of Brown's Valley will assume a more active attitude as soon as the large canal that is being taken out of the Yuba river and is to come to that camp has been completed.

In the Smartsville mines the gravel channel runs north, nearly parallel with the present Yuba river. The bedrock is very uneven; it is for the most part a trap rock. The lowest part of the channel is not very wide and seems to be a fissure in the bedrock, possibly the top of a quartz vein. The bowlders on the bottom are extremely large; the gravel is a blackish blue cement, with occasional layers between of soft gray sandstone. A number of quartz mines have been worked in different parts of the county, but little is doing at present.

SILVER.

RICH DEPOSITS IN WIDELY SEPARATED SECTIONS.

The Mines of Inyo, Mono and Alpine—The Calico and Other Desert Camps—A Remarkable Silver Mine in Shasta County—How Silver Was First Discovered—Depression of the Industry.

History shows that from the most remote period down to within a very short time the money of the world was silver, and that metal was the universal basis of the measurement of values. Up to the time of the discovery of America it is estimated that the entire amount of the precious metals in Europe was only $3,000,000. Prior to 1803 Gallatin, an acknowledged authority on such matters, says that the total product of the mines of the new world, including the amount exported to Europe and that retained in America, was $5,600,000,000. The same authority estimated the product of the American mines in full, from 1804 to 1830, at $750,000,000; the product of Siberia at $100,000,000; that of African gold coast at $450,000,000; of the mines of Europe, $300,000,000 a year. If we add to this product the $300,000,000 existing in Europe prior to the discovery of America, we find a grand total of $7,200,000,000. Allowing for the loss by friction and accidents, imperfect though the data of information be, Mr. Gallatin thought it safe to affirm in 1830 that it certainly exceeded four thousand and fell short of five thousand millions. Of the medium, or $4,500,000,000, he held that one-third to two-fifths was used as currency, and the residue to consist of plate, jewels and other manufactured articles. Of the $4,500,000,000, the presumed then remaining amount in gold and

...ver, he considered that the proportion of gold was considerably greater on account of the exportation to India and China having been exclusively in silver, and of the greater care in preventing every possible waste in an article so valuable. "As the regularity of the annual supply of the precious metals is not," said Mr. Gallatin, "affected by the seasons, the changes in the amount of the last two centuries (preceding 1830) were general and hardly sensible year by year; but the change which took place from 1810 to 1830 was greater than had been experienced since the revolution in values caused by the discovery of America." The annual supply of the mines of America, Asia and Europe had reached its highest point between 1808 and 1810, and amounted then to $50,000,000, or to about 1¼ per cent of the whole quantity of precious metals then existing in Europe and America. The revolts against the Spanish rule in South America broke up the uniformity of the supply and reduced the total annual output to about $27,000,000, or, say, 3-5 per cent of the total amount existing in 1830. Stating in a tabular form the product of America as given by Humboldt, we find the yearly product to have been: 1492 to 1500, an average of $250,000; 1500 to 1545, an average of $3,000,000; 1545 to 1600, an average of $11,000,000; 1600 to 1700, an average of $16,000,000; 1700 to 1750, an average of $25,500,000; 1750 to 1803, an average of $36,300,000; and later by Gallatin: 1803 to 1810, an average of $50,000,000; 1810 to 1830, an average of $27,000,000.

Mr. Gallatin died in 1848, on the very eve of the discovery of the goldfields of California, which occurred in January, 1849. In 1830 Mr. Gallatin wrote, "Specie is a foreign product," that is not native in quantities to American soil.

Chevalier tells us that in the 366 years from 1492 to 1858 there had been produced £401,580 sterling value of pure gold. And that then (1858) Europe was receiving in a single year one-tenth of the total amount sent from the departure of Columbus to 1848. The effect of this great output on the relative value of gold as compared with silver was, however, broken by the interposition of France, which, opening her doors to gold, passed from a silver to a gold standard, and acted as a parachute to break the fall of gold. "Silver," Humboldt says, "had for 2000 years moved from west to east." He added, "The precious metals move in the opposite direction to civilization." In this conversion France was greatly aided by a sudden demand for silver from the far East, so that the Peninsular and Oriental line transported in the year 1856 £12,118,985 (over $60,000,000), and in the year 1857 £16,795,232 (nearly $84,000,000) against £17,000 in 1851. From 1851 to 1861 £110,000,000 value of silver passed to India by the Isthmus of Suez. Chevalier was greatly alarmed at this policy, and feared the consequences to France of a fall in gold should the supply continue. To-day the alarm is in the other direction, a consequence of the vast amount of silver since discovered in Nevada and Colorado.

From 1851 to 1871 £500,000 and $12,500,000,000, an amount of gold equal to the entire existing stock, was added to the world's treasure.

Up to 1870 the production of silver hardly varied from £8,000,000 to £9,000,000 value annually ($40,000,000 to $45,000,000). This was principally supplied by Mexico and Spain, where the silver mines had from early times been of extreme richness. In 1825 they were again worked with great success, and in 1839 the Sierra Almagrera veins began to yield large amounts annually. In 1870 the amount suddenly increased to £15,000,000 value, and so remained for five years, one-half of the amount coming from Nevada alone. Germany, in 1871, and Scandinavia took the alarm, and following the example of France passed to the single standard of gold. The Latin Union limited their common coinage of silver, and England dealt a last and heavy blow by beginning a system of drawing India bills.

Humboldt estimated the entire annual product of Europe, Asia, Russia and the American continent at two millions four hundred and eighty-four thousand pounds, or twelve millions five hundred thousand dollars of gold sent into Christendom, which continued until 1848.

In 1850 the mean sum furnished to the Christian states was £37,950,000 sterling, or $190,000,000. On silver, wrote Chevalier at this time, there had been little change; the product at the beginning of the century had been a little under eight millions of pounds sterling (£7,965,000), and in 1859 was slightly less than nine million pounds sterling (£8,880,000). The falling off in South America occasioned by the political disturbances and the filling of the Potosi mines with vast amounts of water balanced the increase in the product of old Spain.

In the decade from 1871 to 1880, both years inclusive, the gold product of the United States was $346,362,368, an annual average of $34,600,000. In the decade from 1881 to 1890 the gold product was $320,010,725, an annual average of $32,000,000; a total in the twenty years of $666,379,093, an annual average for that period of $33,500,000. To bring these figures to present date, the product of gold in 1891 was $31,555,116. The total product since 1871, twenty-one years, sums up $697,928,209. The fluctuations of product have been from $18,000,000, the lowest point, in 1872, to $48,000,000, the highest point, in 1878.

In the decade from 1871 to 1880, both years inclusive, the silver product of the United States was $184,000,000, an annual average of eighteen and four-tenths millions. In the decade from 1881 to 1890 the silver product was $335,000,000, an annual average of thirty-three and one-half millions; a total in twenty years of $519,000,000. It is hardly worth while to pursue

d asphalt of varying below which, however far sunk, there is um, apparently un-

unk in a number of heavy black oil obntities. Experiment oil is about 90 per and a commencein refining this for nises to become an able industry.

LEUM.

ird Oil-Producing he Union.

the third petroleum the Union, ranking and New York, still here falls far short her of those States. olden State makes a that affords promise it it is not impossible the head of the list. petroleum on this wn for over thirty iorable oil excitetles in Pennsylvania lifornia, and company the score for the oil measures which g the coast all the way reka. A vast amount l in machinery and but the absence of too heavy a handit speedily died out on of the expected n so fondly antici-

rry and disappointthen ten or a dozen g men again stepped ifficulties in the way en removed operawith the result ab ving California the the petroleum proUnion. are confined to the

Deposits by No Means Exhausted—
Abundance of Low-Grade Ore.

The history of mining in Nevada is almost coequal with that of California, gold and silver having both been discovered in that State, then a part of the Territory of Utah, in 1849. In July of that year good placers were found in the ravines tributary to Carson valley, while many of the emigrants who passed through this section in that year en route for the California diggings found gold in different localities, but paid little attention thereto, as they expected to find far richer diggings on the other side of the Sierra Nevada.

Several years passed before any particular attention was paid to the Nevada mines, and it was not until the discovery of the famous Comstock lode that the mining history of Nevada actually began. Some little gold mining had been done at Gold canyon during the first few years after the breaking out of the mining excitement in California, but no one suspected the existence of silver. In 1853 two brothers named Grosch visited Gold canyon and there found ore, which they said they believed to be silver. These men endeavored to raise capital with which to work this ore, but died before being able to do so.

In 1857 gold placers were discovered in Six-mile canyon, a short distance below the site of Virginia City, and among those who took up claims well toward the head of that canyon were two men named Fennimore and Comstock. The first was better known as "Old Virginia," and from these two individuals came the names which were destined to have a worldwide reputation. While searching for gold these miners were frequently bothered by the presence of pieces of some other heavy metallic substance of whose nature they were unaware, and it was not until some one more curious than his fellows took a sample of this metal to Placerville, in California, and had it assayed that the fact was disclosed that it was enormously rich silver ore.

As soon as this became known, which was in the summer of 1859, the famous Washoe rush commenced, and in the space of a month or two a town of upwards of 4000 population had gathered, arastras and then stamp mills were put up in numbers, and some of the great Comstock lode began to yield its millions. At first much of the ore was hauled to California for reduction in the quartz mills of that State, but this soon proved too ex-

GOLD AND SILVER FOUN ABUNDANCE.

Hostile Indians Preventing the D opment of Mines for a Cent More—Rewards Awaiting Ea

It is over a century and a half s first authentic historical accoun given of the discovery of precious in the region now known as Arizon is true tradition from the first adv the Spanish conquerors into Mexi signed to this locality the existen gold and silver mines of fabulous ness, but it was not until 1736 that thing definite was discovered and to the world. In that year a ver silver deposit known as Boles de was found at Arizona, and the I who controlled that region are said have opened some immensely rich mi

But while this section was know possess valuable deposits of gold and ver, its remoteness and the fact th was largely overrun with tribes of and bloodthirsty Indians prevented systematic working or exploratio over a hundred years.

It was not until after the Ga treaty, which gave Arizona and Mexico to the United States, that mines of Arizona commenced to veloped, and even then the Indians so troublesome that the miners lit took their lives in their hands.

In 1855 and 1856 the silver mines Tubac were worked by Americans, as many deposits in the mountains bo ing the Santa Cruz valley. Gold pl were found a year or two later o lower Gila and afterward on the Colo which attracted many prospectors, pa larly to the northwestern part of th ritory.

Many quartz deposits were also and it soon became known that A was blessed with an abundance of th cious metals, but the fear of the A kept the miners from undertaki thing like systematic development for years prospectors were oblige about their work with a pick in o and a gun in the other.

In 1874, however, the Apach conquered and driven from a larg the territory, and at once an era of d opment set in, though handicapped by of transportation facilities and by resul

First American Pig Tin ever produced in Quantity

Shaft House and First Stamp Mill at Cajalco

annual average. When such a difference appears between the outputs of the two decades. To bring the figures to the latest date we add the product of silver for 1891, which is found in the startling figures of $56,000,000, and brings up the grand total to the vast sum of $575,000,000. The change which has taken place in the product of this metal is certainly as extraordinary as that which followed the California discovery of gold. The output rose from one and three-tenths millions in the year 1870 to thirty-two millions in 1880, and fifty-six millions in 1891, without important breaks in the steady progression.

The fiscal year closed June 30, 1891, was startling in its product in the United States. That of gold dropped to a low average point—thirty-one millions; while the silver product leaped over 50 per cent over the largest issue in the twenty-one years stated—namely, from thirty-eight millions in 1888 to fifty-six millions in 1891 (in precise figures, 1889, $37,874,269; in 1891, $56,295,195).

Being so wholly absorbed in searching after and gathering gold, the inhabitants of California paid no attention whatever to silver mining the first decade of the mining era. There were, to be sure, traditions among the native Californians of silver ores having been found in the country prior to its occupation by Americans, Alisal, in Monterey county, being the site of one of these reputed silver fields. As these Alisal ores have since been shown to be poor and scanty, it may fairly be presumed that no argentiferous deposits of any great value were ever met with in California prior to the transfer to the United States and not for about fourteen years thereafter. That our pioneer miners, with so little to encourage them, were not much inclined to hunt after silver so long as the more royal metal continued tolerably plentiful, may well be supposed. Not, therefore, until the discovery of the great Comstock lode, with its great promise of silver, was the attention of our people strongly directed to the business of seeking after and mining for that metal, and even after the occurrence of that event most of the explorations carried on and the mining operations engaged in were for several years conducted outside the limits of this State. It was in the summer of 1861 that Nevada prospectors for silver, who were working south of the Comstock lode, made their way over the line into California. The country first explored by them consisted of the territory at present embraced in Alpine and Mono counties, the still more extensive region lying further south and constituting Inyo county, not having been reached until a year or two later. The silver fields covering these several counties includes all that part of California extending from the summit of the Sierra Nevadas to the eastern boundary of the State, a tract comprising more than 10,000 square miles, its length being 200 and its average breadth fully 50 miles. It is an elevated, rugged, dry and barren region, and, except along the eastern slope of the Sierras, contains very little timber. Not only the higher mountains and their outlying ridges, but also the isolated foothills throughout this tier of counties abound with mineral veins carrying both the precious and the useful metals. While many of these veins contain little ore or only ore of a very low grade, others are powerful, regular and heavily mineralized, carrying large bunches and even considerable bodies of high grade ores.

It was in Inyo county that the first discoveries of silver were made. The region was first visited by prospectors in 1860 who were looking for gold. The country then explored lying to the east, south and west of Owens' lake. Owing to the hostility of the Indians operations had to be suspended and the country was temporarily abandoned. Bishop creek, an agricultural valley, and Owens valley were settled in January, 1862. From that time until 1865 Indian attacks were frequent and it was with great difficulty that the prospecting was carried on. The country proved to be exceedingly rich in mineral deposits. After 1865 the business of mining was successfully carried on at several different points of the county. The most active and largely productive localities since have been the Cerro Gordo, Beveridge, Deep Springs, Darwin, Bishop's creek and Panamint districts. The value of the bullion taken out of these several counties amounts probably to a total of $15,000,000, most of which was the product of the Cerro Gordo mines. For some years little has been done here, though of late there has been a slight revival, and with better prices for silver a large development would certainly occur.

The history of silver mining in Alpine county, where the business has been carried on since 1861 is altogether unfortunate, the product of bullion having apparently been in the inverse ratio of the labor and capital expended on the mines. No portion of the trans-Sierra California is so favorably situated as regards access and facilities for cheap mining and ore reduction as Alpine county. A good level wagon road connects the mining districts with Carson City, fifty miles distant. All supplies for these mines, machinery included, could therefore be obtained during this time at comparatively low rates. The mining districts of Alpine abound with wood and water, a considerable portion of them being covered with large forests, while a number of large streams flow centrally through them, affording much water. The mountains in which the mineral-bearing veins occur are for the most part very steep, making it possible to open up these veins to a great depth by means of comparatively short tunnels. The Director of the Mint spoke thus concerning these mines: "The mining interests of Alpine county still remain, to a great extent, neglected by capitalists, although the showing is of covered a number of years, and the facilities for reaching them are the most favorable. The climate cannot be excelled. A great abundance of wood and water, all these facilities for working our mines cheaply and profitably, seem rather to deter mining men from coming here, and instead, seek investments in more inaccessible and disadvantageous regions."

The most important silver mines in this State are located in San Bernardino county, in what is known as the Calico region. Although silver was known to exist in this region many years ago, it was not until 1881 that a systematic effort was made to work the deposits. In no place heretofore discovered have the precious metals been found under conditions so favorable for extraction. The mineral extends about five miles to the east and west and three or four to the north of the central ore bodies. The pay matter in most instances starts at the surface, and in the shape of gray or greenish chlorides of silver, in vast irregular bodies of decomposed ore at the surface, having a depth of twenty to forty feet, and continuing in those mines most developed in well-defined pipes or chimneys of considerable width to the bottom of the sump, notably in the Silver King, which has reached a depth of over 600 feet, with one wall perfectly developed, and a width of 100 feet. The dolomites of this region are very profitable producers of chlorides and bromides of silver, which are abundantly found, necessitating the addition of pans and wastes to the ordinary stamp mill, and rendering concentration somewhat difficult. The formation includes chloritic schist, highly decomposed, stained by peroxide of iron, in some cases apparently the result of infiltration, and others evidently from the decomposition of sulphuret of iron. Much red jasper interlaminated with crystalline mater is found, particularly on "Wall street." Large quantities of mineralized porphyry giving fine results from pan tests are found in the decomposed strata, running in every direction, from one-quarter of an inch to many feet in width, and yielding $15 to $100 per ton, often running enormously high. So friable is the ore that a single blast will often detach many tons. The surface is quite rolling and everywhere cut by ravines. The ore is often mined by open cuts after the fashion of quarries, as in the Bismarck, Oriental and others of the east group, and can easily be crushed in the hands. Gravity tramways materially reduce the cost of extraction.

Besides the Calico mines silver is also found at many other places in the desert region of San Bernardino county, some of the deposits having been worked successfully as far back as 1861. At present, however, very little work is being done in these mines, the low price of silver and unfavorable legislation preventing capitalists from taking much interest in their development. In a recently published interview a prominent mining expert spoke thus in regard to the ———————: "The rapid decline in the price of silver is having a direct and disastrous effect on one of the principal industries of Southern California, that of silver mining. For years past Calico mining district has been the foremost producer of the white metal in California, but the great depreciation in silver has resulted in the closing of the greatest silver producer in the State, the Waterloo mine.

This bonanza was discovered several years ago and was developed by the Waterloo company to such an extent as to earn the reputation of being the greatest silver mine in California. For years two mills, one of fifteen and one of sixty stamps, had been continuously at work crushing the ores of the great Waterloo properties. Over 200 tons every twenty-four hours were mined and shipped from the vast ore bodies and crushed in the company's mills at Daggett. These operations gave employment to upward of 150 men. The ores, though very free, were low grade, and in the large amount of material handled laid the secret of profit and success. The low price of silver, however, has resulted in the closing down of mines and mills, and to-day Calico and Daggett, both of which towns were recovering from the ravages of fire, are stagnant and almost deserted. The miners have left for other and more prosperous camps, and those who remain are looking forward to a brighter day or are seeking means to migrate to other parts. It has been a great blow to the silver mining industry of the State, and every one interested in mining looks forward anxiously to a rise in the price of silver.

Of the other mines of Calico, those of the Silver King Company are working, and a large number of miners are "chloriding" the richer pockets, for which the Calico district has always been famous. Should the price of silver again go above 90 cents, doubtless the mines now lying idle will resume operations.

Not only is the effect of low-priced silver felt in Calico, but in every other silver mining district in the State. A great many men are leaving for Southern Nevada, or have scattered out over the desert on prospecting trips, and there is little doubt that some of these men, forced by necessity to abandon what they considered permanent positions, will find other mines, and the final result may be more satisfactory than any had hoped for.

One of the most remarkable silver deposits on the coast is located near the town of Shasta, in the county of that name. It is known as the Iron Mountain and has been worked successfully for several years. The mine has a twenty-stamp mill constantly at work, and large shipments of silver ore are regularly made. There is nothing peculiar about the mill, the ore being roasted and worked by what is known as the pan process. The mine itself, however, is a curiosity. There is a solid mountain of ore rising some 1200 feet above the gulch in which the mill is located and extending for miles in either direction. No shafts or tunnels are needed in working the mine, but the ore is simply quarried from the face of the mountain and sent to the mill through chutes. When first discovered this ore was supposed to be a deposit of iron, and so indeed it is, about 75 per cent of it. But assays showed that there was a large proportion of gold and copper. Consequently it was decided to work the ore for the precious metal and let the rest go, and this is now being done.

Underneath the iron and silver combination, however, are immense deposits of sulphurets, which run from $30 to $150 in silver, and which are largely of such a nature that the ore can be shoveled out like so much loose sand. Prospect tunnels have been run in these sulphuret deposits, and no limit to their extent has been found. The mine, had it been discovered on the Comstock or in any other noted mining region, would be one of the wonders of the world, and would produce a second Washoe excitement.

Silver is found in small quantities in other parts of the State, but not sufficiently to repay its working.

The total output of the silver mines of the State up to the present time has not reached $50,000,000, which shows how small a figure the industry cuts, though with proper encouragement it could be made one of the leading sources of wealth.

THE TIN MINES.

California Has the Only Productive Ones in the Country.

Although tin has been known to the world from the earliest times and its value has been fully demonstrated in the arts and sciences, it is one of the rarest of metals so far as deposits that will pay for the working are concerned. The Phœnicians, Greeks, Egyptians and Hebrews made extensive use of tin, and the most ancient writers make frequent reference to the metal. The tin mines of Cornwall have been worked for hundreds of years and there are also deposits in Dover and West Somerset. Tin is found in Altenburg, Saxony; at Nantes, Limoges, Morbihan and Loire Inferieure, in France; in Southern Asia and in Siberia; in Sweden, Spain, in the Malay country, in Madagascar, Australia, Peru and China. It is also found in Greenland and in Mexico.

In the United States the metal has been found in Massachusetts, New Hampshire, New York, New Jersey, North Carolina, Virginia, Dakota, Wyoming, Utah and California.

Outside of the California tin mines those of which the most has been said are at Harney Peak, Dakota. These have been before the public for some time, but the actual results of the development work done do not appear to have been of an encouraging character. The working of several hundred tons of ore developed only about one-half per cent of metallic tin, which is less than the cost of working.

The first American pig tin to be put on the market in any quantity was the product of the mines at Temescal, in San Bernardino county, Cal. These mines are found in a region which resembles geologically in a most marked degree the best tin regions of England. The tin granite formation in which the mineral is found has a known length of twelve miles and a width of four miles. Within this area occur over sixty lodes heavily charged with tin oxide or cassiterite in a remarkably pure form. All the indications point to the fact that these are pure true fissure veins, while shafts and tunnels that have been sunk demonstrate the fact that they extend to an unknown depth.

The history of this tin mine is full of interest. Many years ago, so many that the memory of man nor the written records convey no idea of their number, the Indians who then thickly populated the southern portion of this State found on a hillside in what is now the southwestern portion of San Bernardino county, about forty miles east of Los Angeles, a deposit of jet black mineral, which crumbled on being exposed to the atmosphere and could readily be reduced to powder. In some manner they ascertained, or fancied that they had done so, that this mineral had certain medicinal properties, and it gained wide repute on this account. So the hill came to be called in the native tongue Cajalco, or Medicine hill, and thither came the Serranos and Coahuillas and Dieguitos and Chimehuevas, and even the far-away Yumas, to get store of the wonderful mineral and carry it away with them to their rancherias in the mountains and on the desert.

Very early in the American occupancy of the State the attention of white prospectors, ever on the alert for mineral "signs," was drawn to the strange black deposit on Cajalco. But it puzzled them. It was like nothing they had ever seen. They tested it in the crude methods common in those times, but could make nothing of it, and finally gave up the effort in disgust.

But one day a prospector of more than usual knowledge examined the mysterious mineral, and after several experiments discovered that it was tin ore, the first, too, that had ever been discovered in this country. No sooner was the announcement made than the hills around Cajalco were overrun with prospectors and numerous locations were made. Upward of sixty outcroppings of veins varying from a foot to thirty feet in thickness were found, and two shafts were sunk on a couple of the more promising ledges.

This was a year or two before the war, and while the gold excitement on this coast was at its highest. At once the news of the discovery spread far and wide, and the San Francisco papers contained long accounts of the wonderful tin mines, which it was freely predicted were "worth millions." Great excitement was created, and when some small bars of tin were received in San Francisco there was a great popular outburst, and a second Fraser river excitement seemed imminent.

Unfortunately, however, at the very outset a dispute arose as to the title of the land upon which the mines were located, and the dispute continued for over thirty years, involving tedious and expensive litigation.

This came to an end at last, however, and in anticipation of the final settlement of title preparations were made to commence operations on a large scale. Something over a year ago systematic development of the lodges was undertaken. A branch of the Santa Fe road had been built to South Riverside, a town seven miles distant, which made access to the mines comparatively easy. The work of development was in the hands of a company whose officers reside in England, but which has heavy stockholders in this country. An American was appointed to the superintendency, and under his management extensive improvements were made. Roads were graded, buildings were erected, an immense dam to supply water power was commenced, orchards and gardens were planted. A small five-stamp mill for experimental purposes was erected and two shafts which had been sunk by the discoverers were cleaned out and elaborate hoisting works erected.

On April 25, 1891, the first pig tin was turned out, and in the next two months twelve tons in all of pure tin were produced by the test mill. This mill, it should be understood, was not intended for any but experimental purposes, and was erected solely in order that a thorough test might be made before settling on the site for the permanent and extensive works.

In July the officers of the company became dissatisfied with the slow progress made, and an expert tin miner from Cornwall was put in charge of the works. Prior to this change, however, work had been well under way in the erection of another mill with a capacity of forty tons daily. This was quickly completed and in connection with six concetrators is now turning out from twenty to thirty tons of pig tin monthly. The machinery for another mill of equal capacity is now on the ground and will be put up just as soon as the work can be done, when the production will be doubled.

There are 150 men constantly employed, who are paid the best wages and receive the best treatment as to quarters, food, etc.

So far over 100 distinct veins of tin ore have been found at Cajalco and vicinity, and the ore shows assays of six to sixty per cent, averaging so far ten per cent and showing better the farther down the shafts are sunk. It is pronounced by many disinterested experts who have visited the mines to be the richest ore in the world.

Owing to the formation of the country the ledges are easily worked, both by shaft and tunnel. The ledges all have a slight incline, and those that are being worked widen and become richer as they are worked deeper. There is practically no limit to the amount of ore that can be produced. The only question is that of reduction facilities, and these are being increased as rapidly as possible.

At present crude petroleum is being used as fuel and is found to answer admirably. Tests made of the refuse left in the dust chamber of the furnace after smelting show not a trace of tin—something that does not occur where coal is used.

Including the price paid for the mines and the amount so far expended in their development, over $2,000,000 have been expended, and the payroll amounts to $7000 or $8000 monthly, a goodly share of which is put into circulation at once in the neighboring towns.

Recently extensive improvements have been undertaken, more stamps are being erected and the output of the mines is to be largely increased.

COPPER.

Extensive Deposits in Many Portions of the State.

Copper is one of the most widely diffused metals on the coast, and is found in many localities from one end of the State to the other.

The first copper mines that were worked were discovered at Copperopolis, Calaveras county, on July 4, 1861, by W. K. Reed. This discovery was the cause of great excitement, and one of the most famous "rushes" in the history of mining in California resulted. Copper claims were located on what was supposed to be the extension of this ledge for a distance of fifteen miles. A large town was built up and for a time there was the greatest activity. But the deposits did not repay the anticipations that had been formed, most of the mines were abandoned and for years little or nothing was done.

Within the past few years, however, the two principal mines, the Union and the Keystone, have been reopened, and with favorable results. The ore is shipped East for treatment, and good results are said to have been realized.

The Campo Seco mine, near the village of that name in Calaveras county, has been worked successfully for many years. The ore is very rich and is so accessible that it costs less than $1 a ton to extract. It averages 8 per cent of metallic copper. The mode of working is by roasting in piles of 100 to 1000 tons. About a cord of wood is used for each hundred tons. The roasting is kept up for about four months, after which the ore is passed through a rock breaker, again roasted, this time in furnaces, and then put into the leaching vats, where the metallic copper is precipitated upon pieces of iron.

A notable copper mine is the Newton,

in Amador county. Like locality deposits, mines in this part of the State on a was opened many years ago, and after being worked vigorously was closed down for a long period. It was subsequently reopened and is now said to be paying well. The ore is roasted in heaps in the open air for about six months, some 2000 tons being but into each pile. These are so arranged that there is a good draught through them, and only a small amount of wood is needed to keep the roasting in progress. After the ore has been sufficiently roasted the surface of the pile is sprayed with water so long as any copper solution is formed. Sluices ten feet wide and one foot deep drain the water from the piles, and in these scrap iron is placed, upon which the copper precipitates and is held in the boxes. The amount of copper that results is first dried and then shipped to San Francisco for treatment. Four piles of ore of 200 tons each are kept in operation all the time, and fresh iron has to be put in the sluices every twenty-four hours.

There are large deposits of copper in San Bernardino county, and some of them have been worked, but not to any great extent, because of their distance from transportation and other facilities, being located mostly on the desert in the northern part of the county.

Copper is also found in Alpine, Colusa, Monterey, Contra Costa, Ventura, Inyo, Lake, Mendocino, Sonoma, Mono, San Benito, San Luis Obispo, Santa Clara, Shasta and other portions of the State.

COAL

The Mount Diablo Field—Deposits in Various Localities.

It is over thirty years since coal was first discovered in California, the locality being what are known as the Mount Diablo fields. Since then deposits of this mineral have been found in many parts of the State, and considerable work has been done in exploiting them. Truth compels the statement, however, that none of the coal yet found is of the best type of fuel, though most of it fills the demand where wood is scarce and other coal high-priced.

The Mount Diablo coal fields are situated in Contra Costa county, on the northern and northwestern slope of the mountain of that name, and extend for a distance of twelve miles along its base. The productive portion of the fields, however, but a small part of the entire field, the remainder being so broken and full of faults as to make profitable working difficult if not impossible.

The openings of the leading mines and the dwellings of the workers are concentrated at two villages about a mile apart, known as Nortonville and Somersville. These are located in the bottoms of deep canyons which open out on the San Joaquin plain and which afford easy grades for the short lines of railroad that connect the mines with the shipping points on the San Joaquin river, a short distance above its junction with the Sacramento.

The coal beds which have been profitably worked to a greater or less extent are three in number and are known as the Clark vein, the Little vein and the Black Diamond vein, and they lie in the order named as regards stratification. The total thickness of the entire strata is 359 feet.

The Clark vein is the only one which has been worked continuously through its whole extent or controlled by the company owning it. The coal bed varies in thickness from eighteen or twenty inches to four and a half feet or a trifle more. The average of the entire vein is thirty-two or thirty-three inches. This vein is generally free from slate or dirt of any kind, and makes good, clean coal.

The Little vein is the name given to two contiguous beds of coal, one of which is some fourteen inches thick and the other six inches. In some places these veins reach a thickness of sixteen to twenty-four inches of good coal, and many thousands of tons have been taken out.

The Black Diamond vein varies in thickness in different localities from six or eight to eighteen or twenty feet. But the greater portion of this thickness consists of interstratified clay—slate and "bone" —the last word being a miner's term to designate a very impure, slaty and worthless coal, which forms a weak roof and a bad floor, requiring much timbering and gradually swelling so badly on exposure to the air as to crush the timbers, and necessitate frequent cutting down of the bottoms of the shoots and the gangway floors. The workable coal, wherever it extends in the Black Diamond vein, lies nearly in the middle of the mass forming the thick bed just described, and has bone and shale both above and below it. It generally attains its maximum thickness at those localities where the whole bed reaches its maximum development, or, in other words, where the workable coal is thickest, there, also, the "bone" and slate are thickest, both above and beneath it; and vice versa, where the total tickness of the bed is least, there the workable coal thins out or even disappears entirely, and the whole bed becomes worthless. The coal itself, however, in this bed, wherever thick enough to be worked with profit, is generally clean and free from interstratified slate or "bone," and there have been considerable areas in the Black Diamond vein which have yielded rather better, because harder, coal than most of that produced by the Clark vein.

Throughout the whole length of the upper Black Diamond gangway, except for a little distance in the extreme western portion of the mine, the coal was good, and its thickness averaged about forty-four inches, though varying at different points from thirty-six to fifty-four.

Throughout the Mount Diablo coal mines the beds are frequently more or less disturbed by faults and dislocations. Within the two and a half miles of profitable working, some seven or eight of these faults are of considerable magnitude, involving throws of from ten to fifteen feet to 150 feet or more, and immediately outside of this two miles and a half, both on the east and on the west, there are disturbances of still greater magnitude. But besides these larger faults, the smaller disturbances scattered throughout the mines and involving well marked dislocations, or throws, of from five or six feet down to as many inches or less, are extremely numerous. These disturbances are generally most sharply defined, and may be most easily studied in the Clark vein. Many of the smaller ones are entirely local in character, and extend but very short distances; and it is only a very few of the largest ones which appear to extend through the whole mass of strata between the Clark and Black Diamond veins with sufficient uniformity in character and direction to render it possible to recognize with certainty the same fault in both the veins.

The cost of putting the Mount Diablo coal upon the market has averaged $5 75 a ton, although it is claimed that for some time past coal has been produced here and sold at a profit for considerably less.

The production of these mines reached its maximum in 1874, the total output for that year having been 215,352 tons. Since then the production of coal has fallen off materially, though large quantities are still turned out, which are extensively used in the manufacturing establishments of Stockton and elsewhere.

What is probably a continuation of the Mount Diablo field is found near Livermore, Alameda county, where considerable work has been done at different times. At Corral Hollow three tunnels were driven into the coal bearing formation many years ago. Two of these tunnels were on the "Eureka bed," one of them being about 135 feet higher than the other. The upper one is said to have been 745 feet and the lower one 750 feet long. At the mouth of the lower tunnel a slope or inclined shaft was also sunk some sixty or seventy feet deeper on the bed, which here dips 55 to 60 degrees to the north.

This bed is not well exposed here at the present time. Yet it can be seen that its total thickness is some twelve or fourteen feet, the upper four or five feet of which is pretty clean coal, which can easily be mined separately by itself if desired. The lower portion of the bed is more or less interstratified with slate and dirt, but, in what proportions it is impossible to tell from the existing exposures. Those best acquainted with these old works assert that the bed contains altogether about ten feet of coal, but how much of this it will pay to mine can only be determined by further exploration.

The third tunnel, said to have been 446 feet long, is on what is believed to be the same as the "Livermore Bed," on which a slope was sunk some years ago to a depth of nearly 400 feet on the northeast quarter of section 27, about a mile and a quarter farther west. The strike and dip of this bed are very nearly parallel with those of the "Eureka Bed." The total thickness of the carbonaceous outcrop is six to seven feet, the lower two and one-half feet of which, as now exposed, is clean, good coal. But the present openings are very shallow and it is not improbable that in going deeper the good coal may be found to be somewhat thicker.

Several other beds of coal have been found in this same locality, and in 1890 work was commenced upon a lengthy tunnel which was expected to strike what is known as the Summit bed, but for some reason the project was abandoned after a large amount of money had been expended. It is the belief of experts, however, that there are millions of tons of good coal in this locality, which could be mined and delivered at tide water at an average of not to exceed $2 50 a ton.

Coal has been known to exist at various localities in Humboldt county for over twenty years. The abundance of wood throughout the county has, however, prevented any development of the coal veins. The magnificent redwood forests have as yet been scarcely touched, and outside of the redwood belt there is no lack of oak and other trees, furnishing the best firewood. With refuse wood from the mills at about $1 a cord at Eureka, and given away or burnt elsewhere to dispose of it, and cordwood at $3, coal is used by very few. As return freights for the lumber vessels are always in demand, the amount of coal actually needed at Eureka and the immediate neighborhood is readily supplied at cheap rates. Nevertheless, the time will come when Humboldt's stately forests will be leveled and the available wood burnt. It is then that the more concentrated mineral fuel will be in demand, and the coal deposits will receive due attention. The blacksmiths at various interior points have taken out coal sufficient for their purposes, at the cost of considerable time and labor, rather than pay the enormous freight charges on the imported coal. The advent of a railroad, either from the Sacramento valley or from Ukiah into Humboldt county, will necessarily add a great stimulus to the development of the coal veins. The finds of coal in the county have been quite numerous, particularly in the southern half. Among the localities in which mineral fuel has been found in Humboldt county may be mentioned: The city of Eureka; on Maple creek, three miles from Mad river; two miles north of Arcata, half a mile from the Jolly Giant mill; on the Upper Mattole, on Thomas Rudolph's place; on the main Eel river, two miles below Alder point, on William Wood's place; on Jacoby creek; on Larribee creek; across the Eel river from Eagle prairie, in the bluff; on the Van Dusen, three or four miles above Bridgeville; on the Van Dusen, opposite the Cooper place; on the south fork of Eel river, one mile north of Garberville; on Bear creek, one mile east of Garberville; on Panther gulch and on Buckmountain gulch, tributaries of the east branch of the south fork of Eel river; on the east branch of the south fork of Eel river, on the Ray ranch, and on the Hoopa Indian reservation.

The deposits found upon Eel river have had more or less work done upon them at one time and another, and great things are expected of them whenever suitable railway facilities shall have been supplied.

In Ione valley, at the western edge of the foothills of the Sierra Nevadas, in Amador county, there is a coal bed which has attracted some attention.

This coal is also of very recent origin; quite probably, indeed, not older than some of the auriferous gravels themselves. The bed lies nearly horizontal, and ranges at different points from five or six to twelve or fifteen feet in thickness. It is overlaid and underlaid by a very soft clay rock, and its depth beneath the surface of the ground is small, being sometimes not more than thirty or forty feet. The material itself is strictly a lignite, still showing a good deal of the woody texture. It is not black nor lustrous, but of a dull earthy brown color, very soft and friable, and makes a large quantity of ash. Nevertheless it burns very freely with a bright flame, and the ashes do not form any troublesome clinker. It has been employed for years as fuel for a flouring mill at Ione, the distance to haul it being about three-quarters of a mile.

At the village of Lincoln, in the Sacramento valley, in the southwestern part of Placer county, there is also a coal deposit, of which great expectations have from time to time been entertained.

At American canyon, in the southwestern part of Solano county, there are for some distance in the bluff along the right bank of the canyon heavy but irregular croppings of black carbonaceous shale, containing streaks from one inch to ten inches in thickness of coal. Most of these croppings, however, are not in place, as there has been more or less landsliding nearly all the way along the steep face of the bluff.

The attempt has been made once or twice to organize a company to mine here for coal. But there has never yet been sufficient work done to prove what lies in the solid hill back of the croppings. The locality would also be rather an expensive one to prospect satisfactorily, and the surface indications are not, on the whole, particularly promising. With reference, however, to transportation and proximity to market, the situation is a very favorable one if ever a good mine be found here.

In Orange county, at a locality about twelve or thirteen miles easterly from the town of Anaheim, in the mountains on the south side of the Santa Ana river, not over a mile from the river, and at an altitude of some 1400 or 1500 feet from its bed, there are exposed in the precipitous mountain side some ten or twelve thin seams of impure coal distributed through something like 100 feet in thickness of shales and sandstones, no single coal seam being over about one foot thick.

Some very promising deposits of coal have been found in Fresno county, and considerable development has been done. Four miles northwest of the town of Coalinga is located the mine of the San Joaquin Valley Coal Company. The mine is opened by a series of tunnels. The tunnel through which work is now being prosecuted is 1050 feet in length, and is a crosscut until the main vein is reached. In it the formation is regular in its strata of sandstones, clays and clay shales.

Several small veins are encountered before the main vein is reached, they all being parallel with it. The main vein courses north 20 degrees west, and pitches east at an angle of 30 degrees, and has an average width of four feet. The stratum immediately on the hanging wall of the main coal vein is a compact clay, colored almost black with bituminous matter, and only lacks the luster which distinguishes it from the vein itself. This matter has an almost uniform thickness of five feet. On the foot wall is a soft sandstone six feet thick stratified in itself with thin strata of carbonaceous matter.

About one mile to the east, in section 26, at an altitude of 1000 feet, is the mine of the California Coal Mining Company, four miles from Coalinga by wagon road. The vein here courses north 15 degrees west and dips to the east 35 degrees. It averages two feet in width. The hanging wall is sandstone of an aeranaceous character and the foot wall is clay. The mine has been opened by a tunnel 525 feet in length, running entirely on the vein, giving a vertical depth from the surface at its face of 230 feet. Coal is also found in several other localities in the same region.

There are minor deposits of this mineral in San Joaquin, San Mateo, Santa Clara, Solano, Trinity, Shasta and other sections, but none have been worked to any extent.

QUICKSILVER.

A Widely Diffused Metal and Many Paying Deposits.

Considering the time during which the quicksilver deposits of California have been opened, they rank the first in the world in the amount of production. In

ONE OF COLORADO'S LARGEST SILVER MINES.

It is owing to their immense productiveness that the price of this valuable metal has been reduced to such a point that many mines in this State have been shut down because of the lack of profit in working them.

The commercial status of quicksilver is peculiar. It appears to be three or more times as abundant in nature as silver, and since 1850, according to the reports of the United States Geological Survey, the weight of silver extracted has been about six-tenths that of quicksilver, but the total value of the latter is less than one-sixteenth that of the former metal. This is due to the limited demand for mercury, which is employed in large quantities only for amalgamating the gold and silver ores and for the manufacture of vermilion. Five regions in the world are yielding, or have yielded, great quantities of this metal. These are Almaden, in Spain; Idraі, in Austria; Kwei-Chan, in China; Huancavelica, in Peru, and the Coast Range of California. In the period from 1850 to 1886 Idria produced in round numbers 300,000 flasks of the metal, Almaden 1,140,000 and California 1,400,000, or nearly half the entire product of the world.

As to the time of the discovery of quicksilver in California there is considerable dispute. That the natives of this country used cinnabar in the preparation of paints hundreds of years before the advent of the white man is conclusively demonstrated by the existence of the prehistoric rock paintings at various localities on the coast, into which the hue of red enters more largely than any other color.

As early as 1824 the cinnabar deposit of New Almaden was known to the Spaniards, and in that year an effort was made to extract silver from the ore, those who carried out the experiment not knowing the real character of the deposit with which they were dealing.

In 1835 another attempt of a similar character was made with the same result. In 1845, however, a Mexican army officer who chanced to visit the Santa Clara valley was shown some of the ore and experimented with it in order to learn its real character. With the assistance of one of the padres it was shown that the mysterious metal was quicksilver. From that time on the work of development was pushed until the mines at New Almaden became the most notable in the world with a single exception. These mines are situated near the western limits of Santa Clara county, in a canyon of the inner Coast Range, twelve miles southwest of San Jose. Millions of dollars have been expended here, and many more millions of dollars' worth of quicksilver have been taken out in return; the total yield of this metal for the last quarter of a century being 70,000,000 pounds. The first discovery of quicksilver on this coast was made at Almaden. The natives had used the red ore for paint, but without any knowledge of its mineral character. The veins that have been worked here continuously for over forty years have been nearly or quite exhausted, and it has not yet been determined whether others exist in extent that will repay working.

Next in importance are the New Idria mines, situated in the southeastern part of San Benito county. These mines were discovered in 1852 or 1853, and in 1854-55 active development was commenced and was continued for many years at a large profit. This mine gave rise to one of the most notable cases of litigation that has been known in the mining history of this country. The low price of quicksilver that has prevailed for a long period caused these mines in common with many others to be shut down.

Quicksilver was discovered in San Luis Obispo county in 1872 by a Mexican, in the mountains west of San Simeon, although it was long known to exist in the county by the Indians, who used it as a paint, and were in the habit of visiting the Santa Lucia range of mountains to procure it for that purpose. Over 150 quicksilver claims are recorded in the San Simeon district. In 1871 discoveries of cinnabar were made at Cambria, also about eight miles north of the first discovery, near the northeast corner of Piedra Blanca rancho, which led to the discovery of the Pine mountain lode on the summit of the Santa Lucia. On this lode eight claims were located, from which a large quantity of ore, stated to average 2½ per cent, has been extracted. The Gioson and Phillips claims, the Santa Maria, Buckeye and Jeff Davis, are all located on the same lode. The San Jose mines were located in 1872 upon the eastern slope of the Santa Lucia range. The principal mine that has been developed is the Oceanic. The original claims three in number, were located in 1874, and are situated on the north side and three quarters of a mile from the Santa Rosa creek and five miles from Cambria. The ledge runs east and west, dipping to the north at an angle of about seventeen degrees. The vein is said to vary from eight feet to thirty-two feet in width. At times over 300 men were employed in these works. Three furnaces were erected at a cost of $90,000. Good returns were made on the capital while the price of quicksilver was high, but when it fell to 40 cents per pound it was found impossible to produce it at a profit, and work was suspended.

The quicksilver deposits of Lake county are among the best in the State. Of the mines that have been opened the Bradford is the most important one, and is located a few miles from Middletown, on the Calistoga road. The Great Western mine, which has been worked since 1856, is located four miles south of Middletown. The claim covers 6000 linear feet on the vein, which strikes east and west and dips to the south at an angle of 65 degrees. The hanging wall is clay slate, quite soft near the vein; the foot wall is serpentine. In the first instance the mine was opened by and worked through a tunnel 2200 feet long, intersecting the vein at a depth of 219 feet.

The Sulphur Bank mine is located on the border of Clear lake, ten miles north from the town of Lower Lake. It has been worked for a good many years. Owing to the presence of sulphurous fumes ore extraction is not carried to any great depth. The work of exploitation consists of open cuts and short tunnels.

The Great Eastern mine in Sonoma county has yielded considerable quicksilver, while there are mines that are either now, or were recently, producing the metal in Lake, Merced, Napa, San Benito, Santa Clara, Sonoma, Siskiyou and Trinity counties.

The yield of quicksilver in California from 1850, when the New Almaden mines first commenced to cut any figure in the market, down to the present time has been as follows:

Year	Flasks	Averge Price	Approximate Valuation
1850	7,723	$99 45	$768,000 00
1851	27,779	68 92	1,859,000 00
1852	20,000	58 32	1,166,500 00
1853	22,284	55 45	1,235,500 00
1854	30,004	55 45	1,665,500 00
1855	33,000	53 55	1,768,000 00
1856	30,000	51 65	1,519,500 00
1857	28,204	49 72	1,402,000 00
1858	31,000	47 82	1,482,500 00
1859	13,000	63 12	820,500 00
1860	10,000	53 55	545,500 00
1861	35,000	42 10	1,473,500 00
1862	42,000	36 35	1,526,500 00
1863	40,531	42 07	1,705,000 00
1864	47,489	45 90	1,761,500 00
1865	53,000	45 90	2,433,000 00
1866	46,550	51 62	2,403,000 00
1867	47,000	45 90	2,157,000 00
1868	47,728	45 90	2,191,000 00
1869	33,811	45 90	1,552,000 00
1870	30,077	57 37	1,725,500 00
1871	31,686	63 10	1,999,500 00
1872	31,621	65 07	2,086,000 00
1873	27,642	80 32	2,226,500 00
1874	27,756	105 17	2,919,000 00
1875	50,250	54 15	2,721,000 00
1876	75,074	44 00	3,303,000 00
1877	79,396	38 30	3,041,000 00
1878	63,980	32 90	2,101,500 00
1879	73 684	29 85	2,199,500 00
1880	59,926	31 00	1,860,000 00
1881	60,851	29 80	1,810,000 00
1882	52,732	28 25	1,500,000 00
1883	46,725	27 25	1,275,000 00
1884	31,913	30 50	975,000 00
1885	32,073	30 25	1,060,000 00
1886	29,981	35 50	970,000 00
1887	33,760	42 25	1,425,000 00
1888	33,250	42 50	1,415,000 00
1889	26,464	45 00	1,190,500 00

ASPHALTUM.

Deposits That Promise to Be Sources of Wealth.

Under the general name of asphaltum are grouped a number of bituminous products, including gilsonite, elaterite, rimlite, wurtzilite, albertite, granamite, asphaltum, maltha, brea and bituminous rock. Small deposits of these minerals are found in Utah and Kentucky, but California is the center of production and has by far the largest supply. The total production for the census year 1889-90 for the entire country was some 52,000 tons (since largely increased), and of this California produced 48,000 tons. There is a constantly increasing demand for asphaltum, as new uses are being found for it, and production in California has been greatly stimulated. Much of the asphaltum produced here is of far greater purity and higher grade than that obtained in the island of Trinidad, which is and has for years been the principal source of the world's supply.

Although the greatest use for asphaltum is in the manufacture of street paving, it is by no means confined to that field. Large quantities are consumed in making floors for warehouses, cellars, wineries, breweries, etc. It renders the floors absolutely water tight and is not affected by acids or gases. For lining dams, levees and reservoirs a thin coat of asphaltum put on in a melted state presents a permanent water-tight surface, preventing loss by seepage even when backed by only an earth embankment. As a coating for piling, wharf timbers, ground ends of telegraph poles, etc., it gives almost absolute protection against not only the action of air and water, but also the destructive work of insects and barnacles. It is used as a cement for seawalls and other marine architecture, where its waterproof character makes it especially valuable as a binding material. It is claimed to make wood conduits almost, if not quite as durable as iron, and any iron or other metal work, such as anchors, etc., coated with it will not rust or be affected by sea water. It is also used as a roofing material, and, being practically a nonconductor of electricity, serves a useful purpose as an insulator for electrical wires. Varnish is manufactured from refined asphaltum or gilsonite by simply heating with spirits of turpentine.

The increased demand for asphaltum during the past twenty years is shown by the quantities imported and entered for consumption in the United States during that time.

Years Ended—	Quantity. (Short tons.)	Value.
1867		$6,268
1868	185	5,032
1869	203	10,559
1870	488	13,072
1871	1,301	14,760
1872	1,474	35,533
1873	2,314	38,298
1874	1,183	17,710
1875	1,171	26,006
1876	807	23,818
1877	4,532	36,550
1878	5,476	35,932
1879	8,084	39,635
1880	11,830	87,889
1881	12,883	95,410
1882	15,015	102,698
1883	33,118	149,999
1884	36,078	145,571
1885	18,407	88,037
1886	32,565	108,528
1887	30,808	95,735
1888	36,494	84,045
1889	61,952	138,163

The centers of asphaltum production in California are Santa Barbara, Ventura, Los Angeles and Kern counties, while large quantities of bituminous rock are also obtained in Santa Cruz, Monterey and San Luis Obispo. This latter product is used extensively in street pavements in San Francisco and other cities.

Tests made of the crude as well as refined California asphaltum show that it is superior to the imported, and the table published herewith shows that there is a large field in this country for the use of the products of the California deposits.

An idea of the remarkable features of the asphaltum deposits of this coast can be formed from the following description of the Kern county fields:

"Lying at the very feet of the Coast range, but a little above the valley, yet scarcely within the embrace of the foothills, there is a belt of country in Kern county that at first sight attracts little attention. There are mounds and knolls by means of which the place lies higher than the valley proper and yet seems to have no connection with the mountain system which rises sharply behind it.

"Close investigation shows that these mounds have been built up from the valley at a time not very far back geologically, and by a very peculiar process. The mounds are composed of alternate layers of asphaltum and debris or wash from the mountains. It is evident that at times this asphaltum boiled up and overflowed the banks of the spring, and then during a period of quiet, or by some sudden storm, debris from the hills was deposited, and so on. At one time this region must have been upon a level with the adjacent valley lands, but these mounds of asphaltum have started around the feeding spring and gradually built up to their present dimensions, in a way akin to the formation of geyser crests in the valley of the Yellowstone.

"This formative process is not yet entirely extinct, and one marked and exceedingly interesting example can even now be seen in what are known as the Buena Vista asphaltum fields. Here there are two wells which may properly be called asphaltum 'geysers,' situated about 200 feet apart, but evidently having underground connection, for they pulsate alternately. One has a mouth about five feet in diameter and the other is about three feet across. One is always quiet when the other is in ebullition and the action is like this: Natural gas in forcing its way from down below, will swell the top layer of liquid asphalt until it puffs up like a balloon and finally breaks. After a puff of gas like this from one spring, the surface quiets down and immediately the surface of the other well, 200 feet away, commences to inflate and finally breaks, when the same performance is commenced again at the other end of the line. And so the process goes on, alternating constantly, and from the overflow of asphaltum by this means, both springs are gradually enlarging the mounds around them.

"In this region there are several other bubbling springs of much smaller dimensions, and so far as observed, each acts independently.

"In the Sunset fields, the mounds all appear to have been formed and are now composed of hardened asphalt of varying degrees of thickness, below which, however, in every well thus far sunk, there is found liquid asphaltum, apparently unlimited in supply."

Wells have been sunk in a number of places and a peculiar heavy black oil obtained in large quantities. Experiment has shown that this oil is about 90 per cent pure asphaltum and a commencement has been made in refining this for market, which promises to become an important and profitable industry.

PETROLEUM.

California the Third Oil-Producing State in the Union.

While California is the third petroleum producing State in the Union, ranking next to Pennsylvania and New York, still the amount produced here falls far short of the yield of either of those States. Nevertheless, the Golden State makes a fair showing, and one that affords promise of a development that it is not impossible may in time put it at the head of the list.

The existence of petroleum on this coast has been known for over thirty years. The memorable oil excitement of the early sixties in Pennsylvania had its reflex in California, and companies were organized by the score for the development of the oil measures which were discovered along the coast all the way from San Diego to Eureka. A vast amount of money was invested in machinery and development work, but the absence of railroad facilities was too heavy a handicap and the excitement speedily died out without the realization of the expected fortunes that had been so fondly anticipated.

After this first flurry and disappointment came a lull, and then ten or a dozen years later enterprising men again stepped in, and many of the difficulties in the way of success having been removed operations were resumed with the result already related in giving California the third place among the petroleum producing States of the Union.

The oil measures are confined to the Coast range and its outlying branches, and petroleum has been found in the counties of Los Angeles, Orange, San Bernardino, Ventura, Santa Barbara, Kern, San Luis Obispo, Monterey, Fresno, San Benito, Santa Clara, San Mateo, Alameda,

versity, though younger than the State, always blend its
of this Uni-
age of Cali-
is allied by
heroic times,
who built up
interest by
romance of
, cast their
of this place,
heightened

dearly furnished, at the
The pride of scholarship,
will call forth as much
prizes; but chairs cannot
and equipment held up t
presence of actual want.
How deeply this want

STREET SCENES IN HELENA, MONT.

usa, Humboldt, Shasta and Mendocino. The principal center of production is in the south, the wells of Ventura and Los Angeles counties turning out a constantly increasing quantity of oil.

At Puente, some thirty miles east of Los Angeles, are extensive oil deposits, whose development, however, only dates back to 1882. There are sixteen wells and they produce about 3000 barrels monthly. Most of it is used for fuel and lubricating purposes. These deposits continue into Orange county, where a couple of wells have been sunk near Fullerton which yield small quantities of oil, utilized for fuel. Oil in small amounts is found in other parts of Los Angeles county than the localities mentioned.

The district which yields the largest amount of oil at present is in Ventura county and is a continuation of the Newhall deposits. There are three large companies operating here, and there are wells in Torrey canyon, the Ojai valley and Sepe, Santa Paula, Adams, Wheeler and Aliso canyons. The wells already in existence supply some 800 barrels daily, and new ones are constantly being sunk. There is an extensive system of pipe lines in this territory and a large refinery at Santa Paula. Besides the large companies in operation there are many small wells owned by private parties, and their product is all sold to the large concerns. The crude oil is worth about $1 60 at the well, which is a much larger price than is commanded in the oil regions of the East.

Active preparations have been made for the systematic development of the oil deposit in the southern part of Humboldt county, and the probability is that a large amount of petroleum will be produced here. At present the annual production of the entire State is some 700,000 barrels, but this is capable of almost indefinite extension.

The most recent addition to the petroleum discoveries of the State is in Kern county. Under date of May 11th a dispatch from Bakersfield gave the following news:

"A rich strike of oil in the Sunset fields is an addition to the resources of Kern county that will add millions to its owners. The town of Bakersfield is in a state of excitement over the rich oil in this county. The oil regions of this county for the past year have been persistently and steadily developed. Rich men who had faith in the presence of oil employed experienced talent to develop their property. Well after well has been put down in the Sunset district, each one, until the last discovery, yielding a heavy, black oil, carrying liquid asphalt, and from discoveries already made a profitable industry will be built up. The Southern Pacific Company has already decided to construct a branch line to the fields.

"The company operating in that region, being convinced that what is called 'green oil' could be found, has been persistently hunting for it. Upon striking black oil they would remove the derrick and boring apparatus to another place and try again. The last and successful operation was begun in an entirely new locality a mile distant from former operations, in which they were successful. At a depth of 470 feet green oil entirely free from liquid asphaltum was encountered with a heavy flow of water. The well was put down nearly a hundred feet in this oil-bearing stratum and as soon as the water can be shut off the exact amount of the find can be determined. Other wells are being put down.

"The grand fact of the strike is the quality of the oil. It is of a dark-green color, of about eighteen degrees gravity, and the tests which have already been made prove it to be the very best of the natural lubricating oils, equal in quality and value to the most famed product of West Virginia. An expert who has been testing it during the past week publicly stated to-day that it is a most remarkable oil, not excelled for lubricating and fuel qualities in any locality. The present, as well as the prospective industries of Kern county, have now cheap fuel at their doors, while the market for lubricating oil of such a quality as this recent strike is as wide as the world."

NEVADA.

THE CENTER OF AMERICAN SILVER PRODUCTION.

Hundreds of Millions of Dollars Extracted From Her Mountains—The Deposits by No Means Exhausted—Abundance of Low-Grade Ore.

The history of mining in Nevada is almost coeval with that of California, gold and silver having both been discovered in that State, then a part of the Territory of Utah, in 1849. In July of that year good placers were found in the ravines tributary to Carson valley, while many of the emigrants who passed through this section in that year en route for the California diggings found gold in different localities, but paid little attention thereto, as they expected to find far richer diggings on the other side of the Sierra Nevada.

Several years passed before any particular attention was paid to the Nevada mines, and it was not until the discovery of the famous Comstock lode that the mining history of Nevada actually began. Some little gold mining had been done at Gold canyon during the first few years after the breaking out of the mining excitement in California, but no one suspected the existence of silver. In 1853 two brothers named Grosch visited Gold canyon and there found ore, which they said they believed to be silver. These men endeavored to raise capital with which to work this ore, but died before being able to do so.

In 1857 gold placers were discovered in Six-mile canyon, a short distance below the site of Virginia City, and among those who took up claims well toward the head of that canyon were two men named Fennimore and Comstock. The first was better known as "Old Virginia," and from these two individuals came the names which were destined to have a world-wide reputation. While searching for gold these miners were frequently bothered by the presence of pieces of some other heavy metallic substance of whose nature they were unaware, and it was not until some one more curious than his fellows took a sample of this metal to Placerville, in California, and had it assayed that the fact was disclosed that it was enormously rich silver ore.

As soon as this became known, which was in the summer of 1859, the famous Washoe rush commenced, and in the space of a month or two a town of upwards of 4000 population had gathered, arastras and then stamp mills were put up in numbers, and some of the great Comstock lode began to yield its millions. At first much of the ore was hauled to California for reduction in the quartz mills of that State, but this soon proved too expensive, and soon steam mills were in operation all along the lode.

Much of the ore at first extracted yielded at the rate of $2000 per ton, and as these results became known the country fairly went wild. At first a large proportion of the precious metal in the ore was lost in the tailings owing to faulty methods of reduction, and it is known that millions of dollars went down the Carson river in those days. Efforts have been made to recover a portion of this wealth, and in some cases handsome profits have been realized.

As the working of the mines progressed to lower depths the handling of the water that was found in abundance became so serious a difficulty that it threatened to stop all work below a certain level. To obviate this the Sutro tunnel was planned and excavated at a cost of over $2,000,000, by which the lower levels of the Comstock were drained and their continuation made possible.

The great richness of the mines when first opened caused a vast amount of litigation, rival claimants by the score springing up for each mine, and millions of dollars were thus wasted. A season of depression followed, but in 1873 the famous "Big Bonanza" was struck and then ensued the famous stock-gambling period which has passed into history. While it lasted fortunes were made and lost in a day, the entire coast was demoralized and millions on millions of dollars were lost and won.

The aggregate yield of the Comstock lode down to the present time has been about $320,000,000. In the same period assessments aggregating over $65,000,000 have been levied, and dividends of $118,000,000 were paid, thus giving a balance to the good of only about $160,000,000, the balance having been expended in costs of operation.

The discovery of the Comstock stimulated prospecting in other portions of the State, and many other silver and gold deposits were found. Among the notable localities where important discoveries were made were the Reese river district, from the mines of which over $20,000,000 was taken.

The Cortes district was another that caused great excitement, as the largest ledge in the State was found here, measuring some 400 feet in width by 18,000 or more in length. In so on years the mines on this ledge yielded $20,000,000, but now little or nothing is being done.

In the southern part of the State the Pioche district was the scene of another rush in 1869-70. Here was located the famous Raymond & Ely mine, but after producing over $20,000,000 most of the mines were shut down, though some have been reopened and promise to do fairly well.

The White Pine rush of 1868-69 ranked next to the original Comstock excitement. A number of good mines were discovered, several towns were built and expensive machinery was put in to develop the deposits and work the ore. But like the other districts, the excitement died out almost as quickly as it arose, and the towns were largely abandoned and the mines shut down. Some little work is still being done, and it is believed by many that the district contains many mines that, under modern methods, will yet become bullion producers.

In other portions of the State gold and silver, as well as less valuable minerals, have been discovered, but at present the mining industry is at a low ebb. The last report of the Director of the Mint gave the total value of the bullion output of the State as $8,553,000, it being surpassed by four other localities—Montana, Colorado, California and Utah.

It by no means follows, however, that the mines of Nevada have been exhausted. On the contrary, the bullion yield is slowly increasing, and the time will doubtless come when this State will again rank well to the front in this respect.

ARIZONA.

GOLD AND SILVER FOUND IN ABUNDANCE.

Hostile Indians Preventing the Development of Mines for a Century or More—Rewards Awaiting Enterprise.

It is over a century and a half since the first authentic historical account was given of the discovery of precious metals in the region now known as Arizona. It is true tradition from the first advent of the Spanish conquerors into Mexico assigned to this locality the existence of gold and silver mines of fabulous richness, but it was not until 1736 that anything definite was discovered and given to the world. In that year a very rich silver deposit known as Boles de Plata was found at Arizona, and the Jesuits who controlled that region are said to have opened some immensely rich mines.

But while this section was known to possess valuable deposits of gold and silver, its remoteness and the fact that it was largely overrun with tribes of cruel and bloodthirsty Indians prevented any systematic working or exploration for over a hundred years.

It was not until after the Gadsden treaty, which gave Arizona and New Mexico to the United States, that the mines of Arizona commenced to be developed, and even then the Indians were so troublesome that the miners literally took their lives in their hands.

In 1855 and 1856 the silver mines near Tubac were worked by Americans, as were many deposits in the mountains bordering the Santa Cruz valley. Gold placers were found a year or two later on the lower Gila and afterward on the Colorado, which attracted many prospectors, particularly to the northwestern part of the Territory.

Many quartz deposits were also found, and it soon became known that Arizona was blessed with an abundance of the precious metals, but the fear of the Apaches kept the miners from undertaking anything like systematic development, and for years prospectors were obliged to go about their work with a pick in one hand and a gun in the other.

In 1874, however, the Apaches were conquered and driven from a large part of the territory, and at once an era of development set in, though handicapped by lack of transportation facilities and by resultant exorbitant prices for all supplies. The advent of the railroad, which reached the Colorado river in 1878 and was subsequently extended entirely across the Territory, with the subsequent construction of a second line through the northern portion of Arizona, greatly stimulated development, and a number of prosperous districts were opened up.

The most famous of these perhaps was the Tombstone, where rich silver mines were discovered in 1878, causing a rush that built up a large town and opened up a number of notable mines, including the Contention, Grand Central and Tombstone. These mines have produced many millions of dollars, and while not now yielding so largely as in the past, yet there are still extensive deposits of lower grade ore that will repay working.

The Quijotos and the Harqua Hala districts have also been the scene of extraordinary "rushes," which have resulted in much disappointment, as usual, but accompanied with the development of several valuable mines that continue to produce bullion in paying quantities.

There are some extensive deposits of copper in Arizona, notably at Bisbee and Clifton, and many millions of dollars' worth of this mineral have been produced.

Nearly one-half of the entire area of Arizona is mineral bearing, and many of the deposits are of phenomenal richness, ore yielding from $1000 to $20,000 a ton being not infrequently found. A metalliferous belt, says a well-known authority, extends from the western border of Mojave county, below the big bend of the Colorado river, trending southeastward to Gila county, thence turning southward to the Mexican boundary. Off the main belt are the Yuma county mines of gold, silver, lead and copper, and in the extreme northeast are extensive fields of bituminous coal of good quality, near which petroleum has been found. This great belt may be divided into four groups of mines, the first those of Mojave county, the number of locations there reaching into the thousands. The second group includes the mines of Southern Yavapia and Northern Maricopa, where not less than 10,000 mines have been located. Farther to the southeast is the third group, extending across the Rio Verde into Gila and Pinal counties, the leading districts being Pioneer and Globe. In Pima and Cochin counties is found the fourth group, lying in the mountain ranges bordering on the Santa Cruz and San Pedro valleys, including the famous Tombstone mines.

The total product of the mines of Arizona since it became the property of the United States has been upward of a hundred million, though, as in the case of California, large quantities of gold dust have been produced of which no record has been kept, it not having entered the regular channels of trade.

OREGON.

PLACER AND QUARTZ DEPOSITS DISCOVERED.

Districts Which Still Yield Well—Gold Claimed to Have Been Discovered in the Eastern Part of the State as Early as 1845.

Oregon has never ranked as a gold-producing State, but nevertheless it possesses deposits of that metal which have in the past been a source of no small revenue. Early in 1850 some Oregonians who had visited the placer mines of California and had returned home found gold in the

Umpqua valley, and in the following year good diggings were discovered on the Rogue river and on Josephine creek, and afterward mines were opened in many other localities. In fifteen counties of this State gold has been mined, the leading localities being the counties of Grant, Baker, Josephine, Union and Jackson. In Jackson and Josephine counties the placers are still worked to a moderate extent. In the last-named section there are several hydraulic mines successfully operated. At Sterling creek, Applegate creek and Uniontown are such workings.

At various points along the Oregon coast are large deposits of black sand in which gold occurs in considerable quantities, and which have been worked at different times. At the mouth of the Coquille river these beds are several feet in thickness, but have a superincumbent bed of unproductive sand, which is a hindrance to their being profitably mined. At the mouth of Rogue river are other black-sand deposits, which were discovered and worked with success in 1852, and at various times since then. Much of the sand contains $8 or $9 a ton, but the methods followed in handling it do not save more than a third of that amount, though returning even then a fair profit.

Although placer mining had been successfully prosecuted in a number of localities it was not until 1860 that quartz deposits were found and an effort made to develop them. Most of the paying quartz deposits in this region are found in pockets, several hundred thousand dollars having been realized in two or three years in Jackson county alone from such mines. In the valley of the Rogue river is a pockety ledge which has been traced a distance of twenty miles and is 250 feet in width. Assays run from $2 to $36 to the ton.

In 1860 quartz veins were found on the Jautfain and Moballa rivers, tributaries of the Willamette. One has been found here that assayed $40 to $120 to the ton.

In Eastern Oregon it was believed that gold had been discovered by the emigrants several years before the California excitement of 1848, but nothing definite as to the actual existence of gold in that region was learned until 1859, when rich diggings were found in the Nez Perce country and subsequently on the John Day and Powder rivers. Several mining towns sprang up, and for a while the gold output was very large. Some mining is carried on, though Chinese have largely monopolized the business.

In 1863 quartz was found in Eastern Oregon, and in 1864 a mine was opened which paid largely for awhile. A number of other quartz deposits have also been worked with varying success, a peculiarity being that while the ore paid when worked by the crude arrastra process the more elaborate stamp mills failed to obtain bullion enough to cover running expenses.

The last available reports credit Oregon with a production of gold of over $1,200,000 a year. The counties contributing to this yield are Baker, Curry, Coos, Douglas, Grant, Jackson, Josephine, Linn, Lane, Malheur, Union, Washington and Wallava. Baker and Union counties contributed the bulk of this bullion, the first named also showing a fair product of silver.

WASHINGTON.

GOLD FIRST DISCOVERED BY THE ABORIGINES.

Recently Opened Mines in Various Localities—Abundant Opportunities for Enterprise and Capital.

As in the case of Oregon, it is believed that gold was discovered in Washington some time prior to the historical Marshall incident at Coloma in 1848. It is related that in 1846 a French-Canadian trapper who visited the Cle-Elum river country for the purpose of trading with the Indians noticed a squaw with gold bars pendant from her ears and roughly fashioned anklets of the same metal about her legs. Upon being questioned the Indians, who had been well treated by the trader, told him where they had obtained the gold and showed him a quantity which they had collected. The trapper washed out a quantity of the metal and returned to his home, where he organized a party for the purpose of instituting a systematic search for gold. The entire party was subsequently massacred by hostile Indians, but the fame of the Cle-Elum placers spread, and more than one man "made a stake" there in later years.

The actual mining history of this region, however, is a matter of comparatively recent date. In 1868 the famous Skagit river placers were discovered, causing one of the "rushes" for which the Pacific coast has been so famous. The State geologist of Washington states in one of his reports that the Skagit placers have yielded as high, on an average, as $28 per cubic yard, and instances have been cited wherein miners have secured three and even five times that amount.

While the Skagit placers may now be said to be well nigh depleted of their valuable contents, it is the fact that to this day on portions of the stream placers are being worked with good results by many miners. On Ruby creek, a tributary of the Skagit river, some fabulously wealthy placers were found, and here, too, placer mining is now being carried on with profitable results.

Possibly the placer deposits in Western Washington now engrossing most attention are those located along the Sultan, Snoqualmie, Stillaguamish and Raging rivers in northern Western Washington. Like those of the Skagit, the placers of the Sultan were discovered years ago, the first gold being taken from gravel on the latter stream as far back as 1869.

Aside from the placer deposits of the Skagit and Sultan rivers are those to be found along the Stillaguamish, Snoqualmie and Raging rivers. On the former are some very rich deposits, and the work done in the fields of the Snoqualmie has been attended with an amount of profit sufficient to warrant miners there in embarking in the undertaking next year on a more extensive scale than ever.

On the eastern side of the Cascade mountains are fourteen different streams, along the banks and in the immediate vicinage of which paying placers have been worked for years.

Placer gold was found in Eastern Washington long antecedent to its discovery in the western half of the State. In fact, gold was found along the streams in the middle, northern and northeastern portions of this State way back to the early fifties, and it is believed the aborigines knew of its existence there at even an earlier period. The placers of O'Sullivan creek and Similkameen river have been known these five and thirty years or more to white men. From these placers hundreds of thousands of dollars have been taken, and Chinese miners are said to be working them with profit to this day.

On the Columbia river, from the Little Dalles, near the British Columbia boundary line, to Pasco, in Franklin county, the placers known to be valuable thirty years ago are being worked with profit to this day, but, in the majority of instances, by Chinese miners.

There are many quartz mines in various portions of the State, some of which have paid largely, and many affording promising openings for the investment of capital.

The portions of Washington in which mining is being actively carried on at present are the counties of Asotin, Kittitas, Lincoln, Okanogan, Stevens, Spokane, Whatcom, Walla Walla, Whitman and Yakima.

UTAH.

GOLD FIRST FOUND BY THE MORMON PIONEERS.

Unsuccessful Attempt to Keep It a Secret—Silver Found In All Parts of the Territory—Some of the Notable Mines.

Of the exact date of the first discovery of gold in Utah there exists no record, since that discovery was followed by no such excitement as characterized a similar event on the other side of the Sierra Nevada. It is certain, however, that the Mormons found gold in the mountains near Salt Lake very soon after they settled in that valley. Their leaders, however, discouraged them from prosecuting the search for the precious metal through fear that were the fact of its existence known the territory would be overrun with Gentiles, and Mormon influence would be at an end.

It is impossible, however, to keep the facts suppressed for any length of time, though it was many years after the first discovery before any mines were opened. In 1863 some of the soldiers stationed at Camp Douglas to keep the Mormons in order discovered rich ore in Brigham canyon. The ore carried silver, gold, and lead, the first named metal predominating. Its discovery brought about the very state of affairs that the Mormon leaders had feared. The Gentiles came in in numbers, and the search for mines was stimulated in all parts of the territory.

In 1872 the famous Ontario mine at Park City was discovered, yielding in the first eleven years of its existence $17,000,000, and still paying large returns. Other notable mines in Utah are the Little Emma, which produced $2,000,000 in the first eighteen months after it was opened; the Flagstaff, which has also yielded in the millions; the Horn Silver and many others.

Good mines are now being worked in all parts of the Territory, while discoveries are continually being made that promise to become of importance. In its review of the mining industry of Utah for the past year the Salt Lake *Tribune* said that that period was one of unusual progress and production in nearly all the mining districts. The great Park City district still keeps in the lead. Extensive developments and improvements have been made. In this district are the famous Ontario, Daly and Anchor mines, as well as many others of less prominence, but still good-paying properties. All the old mines here are looking well, while work has been commenced upon the Dolberg group, of which much is expected.

Tintic is one of the oldest camps in the State, and it has passed through the vicissitudes common to all such places. It took years to find out that the rich surface deposits were not all that was good in the lodes. When these surface deposits were worked down to the pyrites, or "white iron," further sinking was stopped, and it has been the work of the past year or two to demonstrate that there is mineral in paying quantities and qualities below this iron stratum, and many old claims will soon become shippers. In the meantime Tintic is spreading out, through having a new district called North Tintic joined to it. A railroad was built into this district during the year, and the towns of Eureka, Mammoth and Silver have taken a new lease of life. West Tintic, too, has come to the front, and Lewiston, Camp Floyd and other places have felt the spirit of enterprise.

Bingham district is another locality where good mines were found years ago, and these from one cause or another were shut down. It is really surprising to see the number of properties opened anew and the results which follow. Part of this activity is largely due to the fact that mines can now be worked at a profit which to work twelve or fifteen years ago simply meant loss to the operator. Cheaper processes of reduction, lower freight rates and less expense in operating a mine now than then are the factors which have worked this change. But in some instances the reason for shutting down these mines of old was the bad management of the companies, mostly such as were outside of Utah. Bingham was never more prosperous than it has been the past year, and the future is certainly bright for the district. Deep mining has proven that the mineral holds its own with depth, and all the companies are preparing to continue downward. The most remarkable record of the year is in the large number of new mines developed, and which have paid from the grass roots down, especially at the head of Carr fork, a part of the district long neglected.

The situation in the two Cottonwoods was greatly improved. There was some excitement last summer about mineral in the foothills between the two Cottonwoods, and quite a number of locations were made. Several open cuts and short tunnels tapped quartz ledges, but as yet not enough mineral has been found to warrant any excitement. It is thought, however, that when the ledges are cut deep enough to find them in place there may be gold and silver enough to pay for putting in reduction works.

Stockton, Ophir and Dry Canyon made good records last year, and there is a brighter outlook for them the present year.

Among the most interesting discoveries of the year were those in Onaqua range where it is crossed by Johnson's pass, and in some of the other ranges on the way to the Deep Creek discoveries, made subsequent to those most exciting of all developments at Dugway and Fish Springs.

The Deep Creek country, of which the last two named districts are a part, is so extensive, so rich in mineral and such an interesting and profitable a prospective field to be reached by a railway, that the proposed line thither has been the subject of great discussion.

There is another interesting situation in the southern country. The developments made by the Dixie Mining Company in Washington county opened up a great copper mine which paid expenses in shipping copper ore of high percentage, and in smelting at St. George and turning out ninety tons of copper bullion.

The revival of Maryvale district and the organization of Gold Mountain district were two important events of the past year. This happened too late in the season to admit of the output being very large, but there was enough preparation to warrant lively times there next summer and a heavy output of mineral for the present year.

La Plata, at the north, drew hundreds of prospectors after the finding of ore there in August, and this led to important discoveries and the opening of quite a number of mines. The whole country, from Ogden, Brigham City and Logan clear over to the Bear Lake valley, is being prospected and many ledges are found, chiefly of galena and carbonates running low in silver, while on the east side of the range is copper, galena and carbonates of lead.

In some of the districts of Utah the product of ores was cut down quite seriously compared with what it would have been had the prices of silver and lead come up to the ideas of miners. Then almost every camp was retarded to some extent by litigation, which tended to stop production. Vexatious as this is, it is an evidence of the value of some of these mines, since there is much truism in the saying that "a good mine invites a contest of title." Again, many mine-owners lack the capital to work their properties, and not a few mines are leased to persons who do not push work, or they bond to those who hold without development. There is need of more capital in all the districts and there are many good opportunities for capitalists to invest.

The entire mineral output of the territory for 1891 was:

1,836,000 pounds copper at 5½ cents per pound	$100,983 30
6,170,000 pounds refined lead at 4 cents per pound	246,800 00
80,356,528 pounds unrefined lead at $60 per ton	2,410,695 84
8,915,223 ounces fine silver at 98¼ cents per ounce	8,759,206 59
36,160 ounces fine gold at $20 per ounce	723,200 00
Total export value	$12,240,885 73

Computing the gold and silver at their mint valuation and other metals at their value at the seaboard, it would increase the value of the product to $16,198,066 81.

NEW MEXICO.

RICH SILVER MINES IN MANY DISTRICTS.

Beneficial Effects of the Tariff on Ores From Mexico—Stimulating Production in Our Own Country—Opportunities for Profitable Development.

New Mexico is a portion of the region which the Spanish conquerors of the new world believed to possess rich mines of gold and silver. That belief has been fully justified by modern discoveries, while the finding of ancient workings in a number of localities would seem to demonstrate that the Spaniards at all events made a systematic search for the precious metals, and appear in some cases to have been fairly well rewarded. No part of the country is so rich with traditions of hidden mines as this, and the experience of the last ten or fifteen years would seem to bear out the truth of tradition in this respect.

In the early part of the present century placer mines were opened and profitably worked in the vicinity of Santa Fe, and during the period of Mexican independence the mines of this section, worked by the most primitive methods, yielded good returns.

After the American occupation subsequent to the Mexican war little was done, owing to the unsettled state of affairs and the troubles with the Indians. With the subjection of the Indian subsequent to the close of the civil war there was a large immigration into New Mexico, and the systematic search for and development of the mineral wealth of the Territory began.

In Taos and Colfax counties much gold mining, placer, hydraulic and quartz was carried on, while rich deposits of silver were found about Pinos Altos, Silver City, Hillsborough and other localities. In 1880 the transcontinental railroad was constructed across the Territory, and this gave mining a great stimulus. Grant, Sierra and Socorro counties are the leading centers of silver production, and the annual output is large and constantly increasing.

In his last annual report Governor Prince stated that the mining industry throughout the whole Territory had increased in amount and profit. "In every section there is an enlarged development. The beneficial effects of the tariff on lead are seen in all of the camps where an argentiferous galena is the staple ore. Relieved of a degrading competition with the ill-paid labor of Mexico, and protected in the receipt of a fair compensation for their arduous and perilous work, our miners are flourishing, and at the same time every such mine is running to its full capacity. The greatness of the present output is illustrated by the fact that the existing smelters are not able to receive all the ore that is produced. This will probably result in the establishment of a large smelting plant at Cerrillos, which has more natural advantages for that purpose than any other point in the Southwest, and probably than any in the country.

"The prospects of the mining industry in New Mexico were never so bright as at present. This is owing to the intelligent and patriotic action of the last Congress. Our principal mineral product is silver, and the great majority of our mines are of low grade, the ore being an argentiferous galena, carrying ten ounces or less of silver to the ton, but being very rich in lead. For several years during the importation of similar ores from Mexico without the payment of duty, these mines in our territory were necessarily closed, for it was impossible for us to compete in the production of these galena ores with the cheap peon labor of Mexico, while our American miners were receiving from $2 50 to $3 50 per day. Perhaps no plainer illustration of the necessity of a proper tariff in order to protect American wages from being reduced to the level of those received by workmen of much lower grade in a foreign land can be suggested than that presented by lead. On one side of the Rio Grande is the intelligent, self-respecting American miner, accustomed to being well fed, well clothed, and to all the conveniences and many of the luxuries of American life, and with ambition to accumulate and become a mine-owner or otherwise independent himself. On the other side is the unintelligent and unambitious laborer, satisfied with the coarse food and hard living to which he is accustomed and asking for nothing better. To subject the former to direct competition with the latter is to reduce him to the lower level or drive him to some other business. Of course he accepts the latter alternative, and so our mines have been closed. But with the protection afforded by the tariff on lead, all this is changed, and the great low-grade mines of the Magdalenas, Cerrillos, etc., will soon echo to the sound of the pick and employ hundreds of well-paid miners."

Grant county is the largest bullion-producing section in the Territory. The first discovery of the precious metals in the county was in May, 1860, when gold was found at Pinos Altos. About 1870 silver mines near Silver City began to be worked, and from that time the produce of both metals has always been considerable, Georgetown being a very steady and reliable camp. Last year the last-mentioned town produced 367,500 ounces of silver, of which the Mimbres Consolidated Company is credited with 230,000. Other large producers were the Mountain Key with $144,000, the Carlisle $150,000 (nearly all gold), and the Graphic $31,000 in silver and lead. Within a short time very rich ore has been discovered in a new locality near Cook's peak. The Santa Rita copper mine, so celebrated for almost a century, and which is the only mine mentioned by Pike in his passage through the country in 1807, is again being worked to the extent of about 250 tons of copper.

There are no less than eighty-five organized mining districts in the Territory as follows: Elizabethtown, Cimarron, Coyote, Guadalupita, Moreno valley, Rio Hondo, Copper Mountain, Taos, Picuris, Arroyo Hondo, Petaca, Mora, Mineral City, Gold Hill, Rio de la Vaca, Pecos, Glorieta, Cerrillos, San Pedro, Galisteo, Bernalillo, Silver Buttes, Nacimiento, Las Placitas, Tijeras Canyon, Hell Canyon, Mount Taylor, Manzano, La Joya, Ladrones, Spring Hill, Council Rock, Gallinas, Iron Mountain, Pueblo, Magdalena, Socorro, Oscura, Hanson, San Andres, San Cristobal, Apache, Black Range, Cuchilla Negra, Cooney, Caballo, Mountain, Rincon, Jicarilla, White Oaks, Vera Cruz, Nogal, White Mountain, Tula Rosa, Jarilla, San Augustin, Lake Valley, Hillsborough, Animas, Percha, Mimbres, Santa Rita, Lone Mountain, Hanover, Silver Flat, Chloride Flat, Pinos Altos, Burro Mountain, Stein's Peak, Virginia and Shakespeare, Cook's Peak, Victoria,

Florida, Tres Hermanos, Carizalillo, Eureka, San Simon, Rio Grande, Rio Colorado, Las Vegas, Mongollon, Capitan, Santa Fe, New Placers and Old Placers.

MONTANA.

THE LEADING BULLION-PRODUCING STATE.

Richest Silver Mine in the World—Over Four Hundred Millions Produced in Thirty Years by the Mines of This Young State.

Montana occupies to-day the proud position of the principal bullion-producing State in the entire country. Since the placers were first discovered in this region, in 1862, the mines have yielded upward of $450,000,000 in gold and silver bullion, and this amount is being added to at the rate of more than $30,000,000 annually.

Gold is found in a greater number of forms here and is more widely distributed than in any other gold-producing region. No laws seem to govern its presence, and it is as apt to be found in one geological formation as another. Indeed, the supposed laws that govern its presence appear entirely inapplicable here.

While gold placers were first discovered in 1862, it was not until two years later that much attention was attracted. In 1864 rich diggings were found on Last Chance gulch, and the man who discovered them took out a fortune in a short time. The news of his good fortune spread rapidly and within a few months hundreds of men flocked in, and the town of Helena, the present capital of the State, was laid out. During the next two or three years many mines were discovered within a radius of 150 miles of Helena, and an idea of their richness may be formed from the fact that in the latter part of 1866 a single shipment of two and a half tons of solid gold was made from Helena.

Quartz was discovered in the Bannack district in 1862, and it was worked in a primitive sort of fashion, but the commencement of the great quartz mining excitement was not until two years later, when the famous Whitlatch mine was discovered in Last Chance gulch. In less than three years this mine produced over $1,000,000, and its discovery marked the commencement of the mineral development of this section, which has resulted in putting it into the front rank of the mining States.

In the ten years from 1880 to 1890 the mines of Montana produced a full quarter of a billion dollars, nearly one-third of which was disbursed in dividends. At the present time this State possesses the largest copper mine, the largest gold mine and the largest silver mine in the United States. These are the Anaconda, the Drum-Lummond and the Granite Mountain.

The principal mines so far developed are in the counties of Silver Bow, Deer Lodge, Lewis and Clarke, Beaver Head and Madison, but in all portions of the State prospecting is actively progressing and valuable discoveries are being continually made. There are in operation in the State ten gold mills, eighteen silver mills, seven lead smelters, eighty copper smelters and twenty-five concentrators, the combined capacity of which is over 5000 tons daily.

The mine which has made the name of Montana more widely known than any other is the Anaconda. This is one of the greatest mines in the world. It is located on a hill which lies northeast of Butte and is a part of the great copper belt that half circles the city. It was purchased in 1881 by Marcus Daly for $30,000, an amount it can now produce every day in the year. There are two three-compartment shafts—one on each, the Anaconda and St. Lawrence—the former 1000 feet deep, and the latter 680 feet deep, it at that depth being on a level with the 800-foot station of the Anaconda. On all the levels between the third and eighth the two mines are connected, and some stoping has been done, but this is inconsiderable, as compared with the vast field that remains to be worked. There are on the different levels miles of drifts and crosscuts tapping the ore bodies, which vary in size, but which have an average of fifty feet in width for a known distance of 3000 feet. These different levels have been worked from time to time, as the general development required, but no attempt has yet been made to work out any portion of the great mass. It is hard to reconcile this statement with the announcement that the hoisting capacity of the works is 1500 tons daily, and that they are worked to their full capacity, and that amount shipped every day to the smelter; yet such is the fact, and while there must be big holes in the mines, there is far more metal remaining than has yet been touched. The policy of the company is only to mine as much ore as can be utilized at the smelters; yet with the wonderful capacity of the latter there are now on the dump at the mines over 100,000 tons. To give an idea of the amount that is taken out of the mines, it may be mentioned that in a single year there has been shipped from them over 600,000 tons, and that the capacity of the railroad company in transporting it between the mines and smelter, some twenty miles distant, has been unequal to the supply. There are employed in and about the mines more than 800 men, all of whom receive large wages. The smelter is the largest in the world, having a daily capacity of 3500 tons, and employing directly and indirectly some 3000 men.

The Granite Mountain mine, located some twenty miles west of Deer Lodge, is the largest silver mine in the world, having paid an average of over $2,000,000 yearly in dividends since its systematic working was commenced. The property of this company consists of eight lodes, two miles long by 2500 feet in width, and the ore runs from $50 to $100 a ton.

The Drum-Lummond mine is situated at Marysville, near Helena, and is an immense body of low-grade gold quartz. In five years it produced $6,000,000, of which over one-third was paid in dividends to the stockholders.

About twenty-five miles south of Helena is the silver-mining center of Wickes, in whose vicinity are four very promising mining districts, viz.: Boulder, Cataract, Colorado and Vaughn. These districts contain a dozen or more mines whose present resources and systematic development justify placing them in the front rank among the very best properties in the West. The Alta-Montana Company of New York city has been fortunate in securing the best of these. The Alta vein averages from four to twelve feet in thickness, embracing galena, carbonates and true silver ores which assay from $60 to $140 per ton in silver and 40 to 60 per cent lead. The ores now being treated from the various mines show an assay value of from $50 to $90. The ore now uncovered and ready for extraction is estimated to supply the great smelting works in operation for years.

The mines of Beaver Head and Madison counties in Southern Montana are steady and large producers and give promise of a future immediately brilliant, because of their nearness to the Utah and Northern branch of the Union Pacific Railway, which is fairly among them. The advantage they possess because of their accessibility and the ease and economy with which machinery and supplies can be transported to them, or ores carried out, are items which will enter largely into their future history. Both counties abound in rich mineral districts, the quartz mines of Beaver Head county, however, just at present attracting by far the larger share of attention. The first gold mining operations of note in Montana occurred in the fall of 1862 at Bannack in this county, and since then some $5,000,000 worth of placer gold has been produced. There are seven organized mining districts within the bounds of Beaver Head. In these are some of the very best mines in the State, and a number which have yielded small quantities of fabulously rich gold ores.

In Missoula county, in the northwest corner of the State, a grand mineral wealth is also indicated by old and recent developments. In the eastern part of the county is Wallace mining district, in which quartz ledges, rich in gold, silver and copper, are plainly traced to a great length by enormous croppings on the surface. In the western portion of the county, on Nine-mile creek, silver mines are being worked, and are producing ore assaying all the way from 40 to 1800 ounces of silver to the ton.

Northern Montana contains the Bear Paw mines, about which so much excitement prevailed early in 1878, and the Judith mines, which created a similar sensation in 1880. Gulch-mining operations in these districts have thus far proved rather unsatisfactory, but residents of Great Falls claim that developments now progressing there will prove them quite extensive and rich. Excellent silver quartz has been discovered in Barker district, near Great Falls. It is a very heavy galena, carrying twenty to fifty ounces of silver per ton and from 60 to 75 per cent lead.

There is not a miner in the Territory who believes that more than one lead has been found out of every hundred that are in the hills waiting for the prospector's pick to uncover them. Extensive regions lying within the Territory have not been prospected at all. Other great areas are known to contain large bodies of ore, but prospectors are not sufficiently numerous to permit those regions to be prospected. Well educated men who are familiar with the mineral belts of the Territory declare their belief that there is more gold and silver in Montana than there is now in circulation in the world, and that her mines will, inside of the next seventy-five years, yield $5,000,000,000, which sum equals the money in circulation to-day.

IDAHO.

RICH PLACER DEPOSITS AND QUARTZ MINES.

The Wood River, Owyhee and Cœur d'Alene Discoveries — Immensely Rich Gravel Mines and Quartz Ledges—Ore Worth $4 an Ounce.

The first discoveries of mines in Idaho were made while this section was still a portion of Oregon and Washington. In 1860 the discovery of rich placers on the Clearwater and Salmon rivers attracted many prospectors, though trouble with the Indians prevented any great development for several years. In 1862 the famous Boise placer mines were found, in which the average earnings were $18 a day, and $100 to the pan was nothing extraordinary. Several cases are reported where miners took out as high as $1500 a day for longer or shorter periods. Many millions of dollars were washed from these rich placers, the annual production having been upward of $10,000,000 for several years. The palmy days of placer mining in Idaho were from 1860 to 1865.

Then came a period of depression, which lasted for a long time, but about ten years ago there was a revival in quartz mining, which has continued to increase constantly down to the present time. The first great quartz mining excitement in Idaho occurred in the southern part of the then Territory in the summer of 1863. Several deposits were found here which assayed from $1200 to $2700 a ton in gold, accompanied by silver to the value of $27 to $95. On Granite creek other rich discoveries followed, notably one from which the poorest rock gave $62 to the ton and the richest $6000 to $20,000. Many towns sprang up in the Boise basin in consequence of these discoveries and thousands of people poured into the country. The remoteness of the district from transportation lines made the introduction of mining machinery very expensive and, in fact, almost impossible, and for a considerable period the primitive arastra was the sole method of working the rich quartz that was found in abundance. At one time there were eighty-four of these crude mills at work in South Boise.

In 1863 the rich Owyhee placer fields were opened, and at the same time many valuable silver ledges were found in the same region. The first quartz mill in this region was erected at War Eagle mountain in 1864, and in 1865 in the same locality was discovered the famous Poorman mine, the ore in which was the richest ever known. It was chloride of silver, impregnated with gold, and was sold at the rate of $4 an ounce just as it came from the mine.

In 1864 the first discoveries in the Wood river region were made, but owing to Indian and other troubles it was not until 1879 that this section was really opened. The Wood river rush is famous in the annals of mining in the West. Ore that produced $11,000 to the ton was found, while the average ran from $100 to $500. The mineral belt extends some eighteen miles in length, and many mines have been opened and mills erected which are producing millions of dollars annually.

In 1875 the Yankee Fork district, north of Salmon river, was opened, the first discovery there being a mine that yielded $2000 to the ton. Among the notable mines of this district is the Custer, which was largely worked by simply quarrying out the ore from the surface. One of the largest silver veins known in the world is situated here. It is called the Cow's Horn and there are twenty-four claims on it, each of which is 1500 feet in length. The ore assays as high as 800 ounces of silver to the ton.

The Cariboo placers, on the headwaters of the Snake river, were discovered in 1870 and yielded a quarter of a million annually for ten years or more. Subsequently quartz was found which assayed high and has given the district permanence.

One of the notable rushes of which Idaho has been so prolific was the Cœur d'Alene excitement in 1883. That district has gone through the usual history of mining development, and has a number of quartz and hydraulic mines that pay well.

The leading bullion producing section of Idaho at present is Shoshone county, which turns out over $2,000,000 annually, the bulk being silver. Next in rank is Owyhee county, with a product of nearly a million and a half, and then come Custer, Alturas, Boise, Lemhi, Logan, Idaho and Elmore.

There is an almost unlimited opportunity in Idaho for the development of good-paying mining properties. The extension of railroads through the State is opening new districts and opportunities for the profitable investment of capital are to be found in great numbers.

COLORADO.

IMMENSE PRODUCTION OF SILVER AND GOLD.

An Average of Thirty Million Dollars Annually Turned Out of Her Mines—Where Gold Was First Discovered—The Usual Excitement.

Colorado disputes with Montana the honor of leadership in the production of the precious metals, the output of the two States mentioned having been very nearly equal for several years past, being in round numbers some thirty millions of dollars. The figures of Coloado's production for the twenty years between 1870 and 1890 were as follows:

Prior to	Gold.	Silver.	Copper.	Lead.	Total
Prior to 1870.	$27,213,081	$330,000	$40,000		$27,583,081
1870.	2,000,000	650,000	25,000		2,620,000
1871.	2,000,000	1,029,046	30,000		3,039,046
1872.	1,725,000	2,015,000	45,000	$5,000	3,790,000
1873.	1,750,000	2,185,000	65,000	28,000	4,028,000
1874.	2,002,487	3,090,023	90,197	73,676	5,262,383
1875.	2,161,475	3,122,012	90,000	60,000	5,434,387
1876.	2,726,315	3,315,592	70,000	90,000	6,191,907
1877.	3,148,717	3,726,379	93,790	247,400	7,216,283
1878.	3,490,384	6,341,807	89,000	636,924	10,558,116
1879.	3,193,300	15,385,000		532,362	24,500,000
1880.	3,200,500	18,615,000		1,678,500	23,500,000
1881.	3,860,000	16,500,000			22,544,170
1882.	4,100,000	17,370,000			21,470,000
1883.	4,300,000	16,000,000			20,300,000
1884.	4,300,000	19,990,351			24,290,351
1885.	4,446,417	18,200,406			22,655,823
1886.	4,000,000	13,500,000	155,847	5,890,000	23,390,500
1887.	4,123,935	17,121,978	375,328	5,790,000	27,194,197
1888.	4,038,748	10,657,740	359,440	5,089,720	29,941,000
1889.	4,512,136	20,259,901		4,749,852	29,881,334

The history of mining in Colorado is replete with interest, that State having been the scene of some of the most remarkable discoveries that have ever been recorded. When the fact first became known that the precious metals existed in Colorado there is no record to show. In the southern portion of the State, where are the remarkable ruins of the cliff dwellings, have been discovered abandoned shafts and other indications of the discovery of gold or silver, and perhaps both, by prehistoric peoples. By some it is thought that the Spanish conquerors of Mexico penetrated into this region in their wild search for gold, but of this there is no proof.

Coming down to later times it is a matter of tradition that the Indians who frequented the South Park and other localities were possessed of lumps of gold found there by them, and when the great army of gold-hunters poured across the plains to the mines of California there were not wanting an abundance of rumors that the streams and gulches on the eastern slopes of the Rocky mountains contained the precious metal for which all were in search. But the determination to reach the mines of California was too strong to permit of loitering or halting by the way, even though the existence of mines in Colorado had been positively demonstrated, and hence years passed before those mines became known.

It was not until 1858 that the question was definitely settled by the discovery of placer deposits near Cherry creek, and this was followed by an excitement second only to that which followed the discoveries ten years before in California. The famous "Pike's Peak rush" is historical. It was estimated that fully a hundred thousand people poured into this region during the year after the first discovery of gold in Colorado, a large share of whom returned to the East discouraged and disheartened.

Following the discoveries in the vicinity of where Denver is now located, came the location of rich mines in the southern part of the State, it being a remarkable fact, however, that while Colorado was destined to win her greatest fame as a producer of silver, the earliest miners knew nothing of the presence of that metal, but confined their efforts entirely to the search for gold.

In a short time after the first discoveries the yield of gold amounted to $7,500,000 annually, and an unprecedented rush took place which lasted for several years. The placer diggings were thronged, and soon quartz deposits were discovered which proved very rich. In July, 1859, the first arrastra was put in operation, and within a few months several steam and water-power quartz mills were at work.

The same methods that had been employed in California in working gold quartz were put into operation here, but the ores proved refractory, and many mills were shut down because of their inability to save enough of the gold, that was undoubtedly present, to pay expenses. The output of the State fell off materially, and a season of depression ensued.

All this time, while the presence of silver was known to a greater or less extent, no attempt had been made to work the ores, in fact their richness not being understood.

It was not until 1868 that a smelting establishment for handling silver was put into successful operation, and from that time the permanent prosperity of Colorado and its prominence as a silver-producing State is actually dated. There has been a steady increase in the bullion product of the State until it has reached the point at which it has remained for several years—not varying much from $30,000,000 annually.

The most remarkable of all the Colorado mining regions in many respects is Leadville. This section was at first a placer mining camp, and fairly good diggings were found. In digging for gold heavy masses of a peculiar rock were discovered which were cast aside as worthless, nor was it for a long time that it was learned that this despised material was silver ore of the richest kind. The ledges at Leadville are what is known as blanket veins, and from this section comes fully half of the bullion product of the State.

New discoveries are constantly being made in various portions of the State, during the past year much excitement

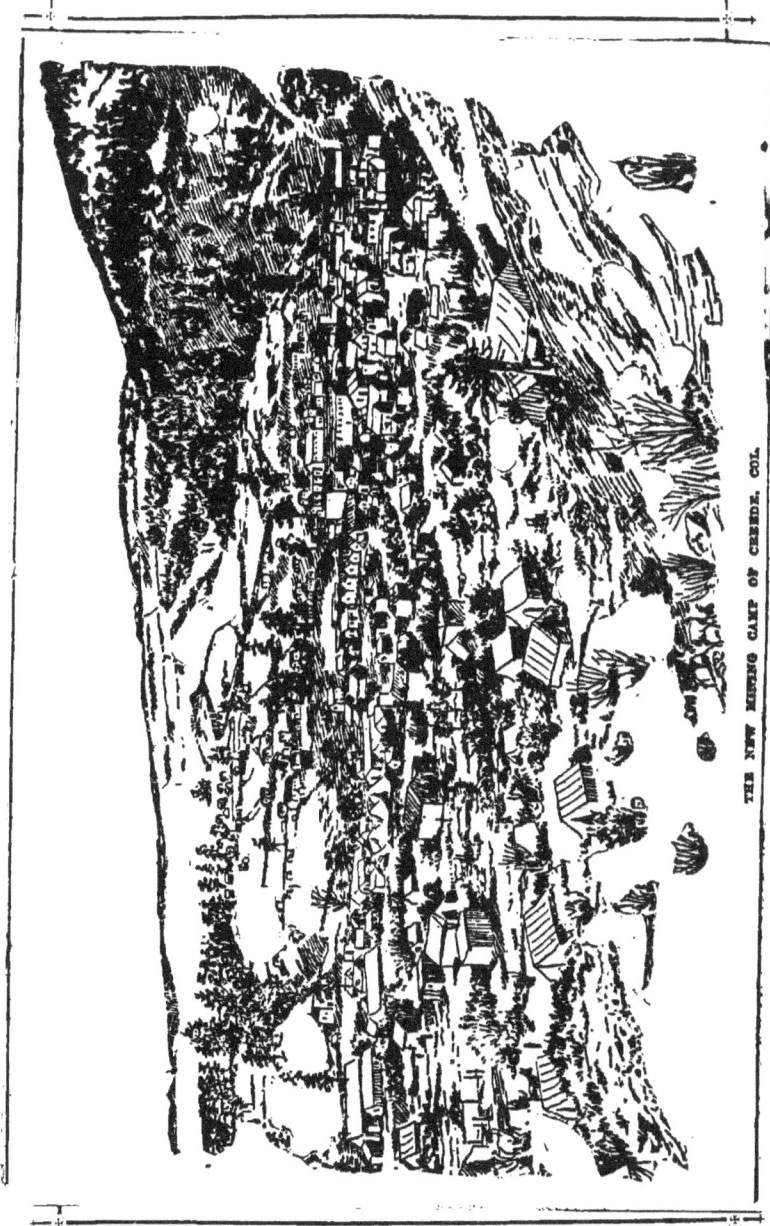

THE NEW MINING CAMP OF CREEDE, COL.

having been caused by the Creede and Cripple creek rushes, great things being promised for those camps.

ALASKA.

The Largest Quartz Mill in the World.

Mining is carried on extensively in Alaska, and prospectors may be found in all parts of the Territory. This region can boast the largest mill in the world. It is situated on Douglas island, at the mine known as the Treadwell, and with its 240 stamps, ninety-six concentrators and twelve ore crushers, forms a veritable sight for tourists. In the Silver Bow basin, near Juneau, and other points in Southeastern Alaska, the quartz mines are being actively developed. There are fifteen mills in operation for crushing the ore brought from the various mining locations within a radius of 500 miles from Sitka. The low-grade ores found in this portion of the Territory yield 66 ounces of silver and $4 in gold to the ton, those of higher grade giving returns of 340 to 350 ounces of silver and $22 in gold to the ton. Considerable quantities of silver ore have been shipped to San Francisco for treatment, with satisfactory results. In the Berner Bay district development has been carried far enough forward to demonstrate to a certainty the fact that the lodes are strong, well defined and unusually rich. Promising silver discoveries are also reported to have been made in the same district, while a vein of that mineral recently opened on an island in Glacier bay is said to be yielding results far beyond the most sanguine anticipations of the owners.

Numerous new discoveries are announced from various parts of Southeastern Alaska. One of these not far from Juneau City promises the addition of a number of valuable mines to those already opened or in progress of development. The discovery, located on one of the branches of Sheep creek, which last empties into Gastinaux channel, nearly opposite the big mill on Douglas island, is believed to be an extension of the Silver Bow basin belt, though in some of the lodes silver predominates, which is not the case with those of the Silver Bow basin. The ore carries galena, zinc blende and copper pyrites, the assays showing well in gold, with a very large percentage of silver.

What appears to be a most promising silver-mining district is that of Golovin bay, or rather of Fish river, a stream which empties into the bay of that name. This silver belt is located in the mountain range of the broad peninsula which projects itself to the westward between Behring sea on the south and the Arctic ocean on the north and in latitude 65 deg. It is about thirty miles distant from the navigable waters of Golovin bay, which is a branch or arm of Norton sound, and the only disadvantage is that the ore has to be packed some miles to the river and thence transported in light-draught boats to the head of ocean navigation. The ore is an argentiferous galena, carrying from 75 to 85 per cent of fine lead and from $100 to $250 in silver to the ton. In the Yukon country a great deal of gravel mining is done, but the seasons are so short that little profit remains to the miners.

MINING LAWS.

How to File Claims Upon and Obtain Patents for Mines.

In connection with the subject which has here been treated at length it will be of interest to many to give a digest of the laws governing the acquisition of mining lands, and the process of "taking up" claims and obtaining patents therefor.

The law governing the location of clims on veins or lodes provides that any person who is a citizen of the United States or who has declared his intention to become a citizen may locate, record and hold a mining claim of 1500 linear feet along the course of any mineral vein or lode subject to location; or an association of persons, severally qualified as above, may make a joint location of such claim of 1500 feet, but in no event can a location of a vein or lode exceed 1500 feet along the course thereof whatever may be the number of persons composing the association.

With regard to the extent of surface ground adjoining a vein or lode, and claimed for the convenient working thereof, the Revised Statutes provide that the lateral extent of locations of veins or lodes shall in no case exceed 300 feet on each side of the middle of the vein at the surface, and that no such surface rights shall be limited by any mining regulations to less than twenty-five feet on each side of the middle of the vein at the surface. Said lateral measurements cannot extend beyond 300 feet on either side of the middle of the vein at the surface, or such distance as is allowed by local laws.

The statutes provide that no lode claim shall be recorded until after the discovery of a vein or lode within the limits of the ground claimed; the object of which provision is evidently to prevent the encumbering of the district mining records with useless locations before sufficient work has been done thereon to determine whether a vein or lode has really been discovered or not.

The claimant should therefore, prior to recording his claim, unless the vein can be traced upon the surface, sink a shaft or run a tunnel or drift to a sufficient depth therein to discover and develop a mineral-bearing vein, lode or crevice; should determine, if possible, the general course of

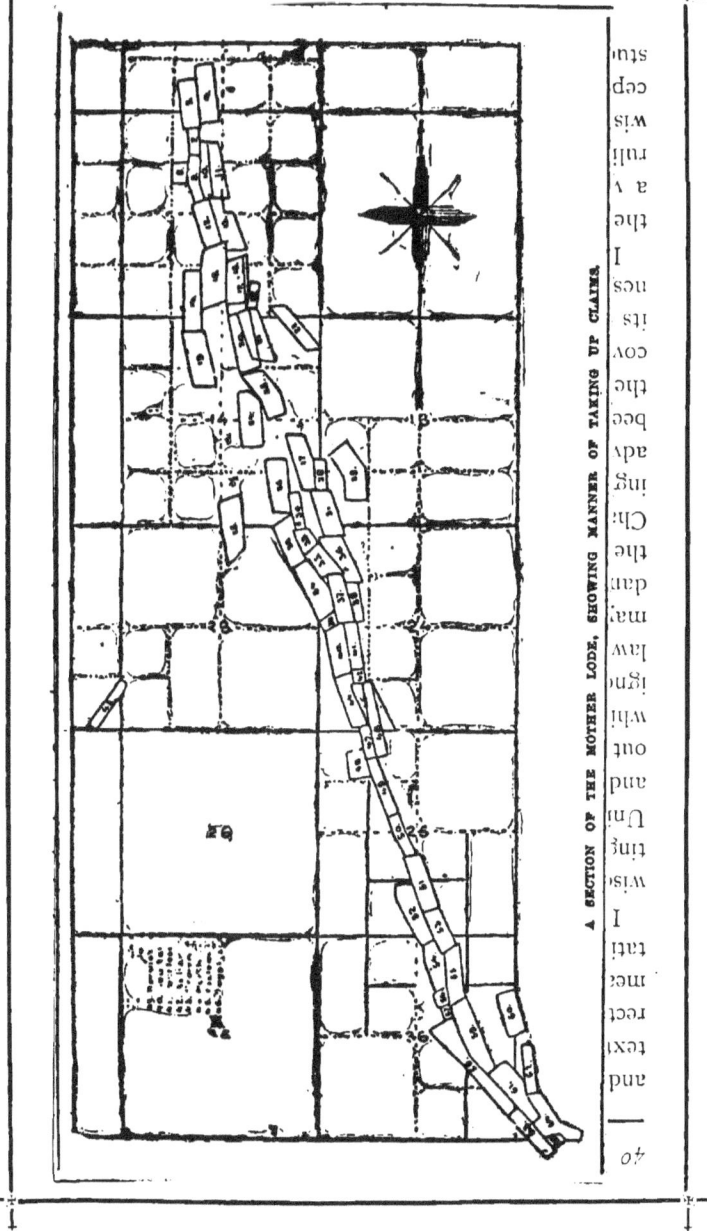

A SECTION OF THE MOTHER LODE, SHOWING MANNER OF TAKING UP CLAIMS.

th vein in either direction from the —int of discovery, by which direction he will be governed in marking the boundaries of his claim on the surface, and should give the course and distance as nearly as practicable from the discovery. shaft on the claim to some permanent well-known points or objects, such, for instance, as stone monuments, blazed trees, the confluence of streams, points of intersection of well-known gulches, ravines or roads, prominent buttes, hills, etc., which may be in the immediate vicinity, and which will serve to perpetuate and fix the locus of the claim and render it susceptible of identification from the description thereof given in the record of locations in the district.

In addition to the foregoing data the claimant should state the names of adjoining claims, or, if none adjoin, the relative positions of the nearest claims; should drive a post or erect a monument of stones at each corner of his surface-ground, and at the point of discovery or discovery shaft should fix a post, stake or board, upon which should be designated the name of the lode, the name or names of the locators, the number of feet claimed, and in which direction from the point of discovery; it being essential that the location notice filed for record, in addition to the foregoing description, should state whether the entire claim of 1500 feet is taken on one side of the point of discovery, or whether it is partly upon one and partly upon the other side thereof, and in the latter case, how many feet are claimed upon each side of such discovery point.

Within a reasonable time, say twenty days after the location shall have been marked on the ground, or such time as is allowed by the local laws, notice thereof, accurately describing the claim in manner aforesaid, should be filed for record with the proper recorder of the district, who will thereupon issue the usual certificate of location.

In order to hold the possessory right to a location not less than one hundred dollars' worth of labor must be performed, or improvements made thereon, within one year from the date of such location, and annually thereafter; in default of which the claim will be subject to relocation by any other party having the necessary qualifications, unless the original locator, his heirs, assigns or legal representatives, have resumed work thereon after such failure and before such relocation.

The expenditures required upon mining claims may be made from the surface or in running a tunnel for the development of such claims, the act of February 11, 1875, providing that where a person or company has or may run a tunnel for the purpose of developing a lode or lodes owned by said person or company, the money so expended in said tunnel shall be taken and considered as expended on said lode or lodes, and such person or company shall not be required to perform work on the surface of said lode or lodes in order to hold the same.

The importance of attending to these details in the matter of location, labor and expenditure will be the more readily perceived when it is understood that a failure to give the subject proper attention may invalidate the claim.

Under the law governing placer claims lands may be taken up which contain all forms of mineral deposits except veins of quartz or other rock in place. These claims must not exceed twenty acres in area for each person, and the method of filing notices and perfecting title is the same as for obtaining patents to lodes.

MINING MAGNATES.

The Advance Guards of the Industry.

Discoverers and Developers of Mines.

The Part They Played in Pacific Coast History—Notable Collection of Names and Faces.

It is but just in this connection to refer to some of those men whose names are indissolubly linked with the history of mining on this coast, and to whose energy, enterprise and unbounded faith in the hidden resources of the ledges and gravel deposits is due the vast additions to the wealth of the world that have been made by the Western States and Territories of this continent. Many of the most prominent of these men have long since "gone over the range," but their memory is still strong in the hearts of the survivors, and will ever be perpetuated. Their names will pass into history and will endure so long as the story of the wonderful discoveries of the Western slope shall be told, which will be for all time. They were a race of giants and so long as California shall be known, so will they be.

JAMES W. MARSHALL.

He Discovered Gold, but Died a Poor Man.

The place of honor in a series of sketches of this character certainly belongs to James W. Marshall, the man who made the discovery that literally set the world ablaze with excitement.

Marshall was born in Hope township, Hunterdon county, N. J., in 1812. His father was a coach and wagon builder, and he was brought up to the same trade. His early life presents no features of special interest, and he had arrived at man's estate—being just 21—when he began to turn his eyes westward. Others of his neighbors were of the same mind, a party was formed, and the start was made May 1, 1844. The way was tedious, but not particularly exciting, and after wintering

James W. Marshall.

in Oregon, Marshall and his friends entered California, via Shasta, in June, 1845, coming down the Sacramento valley. Here the party separated, Marshall going to Sutter's Fort. For several months he pursued the even tenor of the dull life at the fort, stocking plows, making spinning wheels, mending wagons and doing such general carpenter work as was required.

In August, 1847, Sutter started a mill at Coloma, on the American river, placing Marshall in charge of it, and it was through the action of the water which was used as the power of the mill that gold was first discovered in California. Among the debris which accumulated first beyond the mill wheel Marshall saw on that memorable day of January 24, 1848, what he supposed to be pebbles of an unusual color and of great weight. He took them to Sutter, who, after a careful examination and test of them with nitric acid, told him they were gold nuggets. This was the manner in which gold was discovered in California.

The news spread like wildfire. The residents of California started in quest of the precious metal and the great gold fever begun which revolutionized the Pacific coast. But what was other men's fortune was the discoverer's ruin. Marshall engaged in mining with varying fortunes in various sections, but finally drifted back to the scene of his discovery and died there in extreme poverty a few years since.

GENERAL JOHN A. SUTTER.

The Man to Whom Many Californians Owe Everything.

The name of General John A. Sutter was more extensively associated with California than that of any of the pioneers of the new State. He was the son of a Swiss of the Canton Berne, but was himself born in Baden, Germany, on Feb. 28, 1803. His early life was passed in the Grand Duchy, and there he received his education. Like many of his countrymen, he wished to enjoy the freedom of our republic, and sailed for New York, where he landed in July, 1834. Leaving family and home in company with Sir William Drummond Stewart, he joined a party under the charge of Captain Tripps of the "American Farm Company" and started

General John A. Sutter.

for the broad valleys of California, where he knew there was rich soil which only awaited the cultivator to bear unlimited fruits, and where Providence had even a still richer yielding field that he knew not of. He left his party in Kansas and with five companions he started for his destination by way of Oregon. After numerous perilous adventures he finely arrived at Monterey in 1839. Having overcome the Spanish opposition to foreign settlers he obtained permission from Governor Alvarado to locate himself in the valley of the Rio del Sacramento. He explored the Sacramento and American rivers, and about eighteen months after he commenced his journey permanently established himself on the latter river with a colony of three whites and eight Kanakas. Shortly after he removed to what is now known as Sutter's Fort and took possession of the neighboring country under a Mexican grant. In the winter of 1847-48 he had contracted with James W. Marshall to erect a sawmill on the south fork of the American river, and on January 24, 1848, through this mill that great discovery was accidentally made which revolutionized the country. With him, as with its discoverer, the finding of gold seemed rather to deter than to advance his interests, and he died about twelve years ago in comparative obscurity.

GEORGE HEARST.

One of the Foremost and Most Successful Mining Men.

No man has played a more prominent or important part in the mining development of the Pacific coast than the late Senator George Hearst. In his long business career on the Pacific coast he made many friends and few enemies. The California life of George Hearst was full of exhibitions of perseverance, success, reverses and finally a rich harvest which he

Senator Hearst.

reaped from legitimate business operations.

He was born in Franklin county, Missouri, on September 3, 1820. Born and raised in what was then the chief scene of mining operations in the United States it was quite natural that the young Missourian should early enter into the business of mining as a profession. He soon achieved quite a success in the mining of copper and lead in the locality of his home. Attracted by the reports of the "golden fleece" upon the slopes of the western mountains he, in company with several of his neighbors, left their home in March, 1850, and made the trip across the plains in five months, arriving at Placerville, El Dorado county, in October of that year. He immediately started to work at placer mining, with the varied success attendant upon that class of operations. He was sent by a company in 1859 on a prospecting trip to the Comstock lode, then in Utah Territory. At that time very few of the miners knew the real value of silver and most of them mined solely for gold. Some, however, had assays made for silver, and one gave the information to Mr. Hearst that this "black stuff" contained $2000 to the ton in that metal. He remained prospecting for about six weeks and made up his mind that the discovery was of immense importance and that mines of vast wealth would be developed.

He at once made contracts for interests in one of the most important strikes, which is now known as the Ophir, and returned to Nevada City to obtain the required purchase money. Having succeeded, he again journeyed to the Comstock and proceeded to work his claims. In 1865 mining enterprises in which he was engaged proved disastrous and he found himself in 1866 in reduced circumstances. However, he regained his fortune by other mining enterprises, outside of California as well as in this State, and by judicious investments in San Francisco real estate. He was a member of the State Legislature of California in 1865-66, was appointed United States Senator by Governor Stoneman in 1886 to succeed Senator Miller, deceased, and elected to the office in 1887.

He died recently after a lingering illness in Washington, D. C. His name is connected with many of the largest and most successful mining enterprises in the West.

WILLIAM C. RALSTON.

He Rose From a Freight Clerk to a Bank President.

William C. Ralston was born in the pretty little village of Wellsville, O., Jan-

William C. Ralston.

uary 12, 1826. He was the son of the principal builder in the town, and, naturally, when the proper age was reached, the boy was put to learn his father's business. Although he served the required time in the trade, he had made his mind up to earn his living by some other means, and upon graduating from his apprenticeship he secured a position on one of the boats of the Mississippi Steamship Company as freight clerk.

After filling this position for a while he was sent to Nicaragua by C. K. Garrisson, one of the directors of the Mississippi line, to look after interests of his centered in that locality. In 1849 he was sent to California to take charge of an agency for the Vanderbilt Steamship Company, which was established a short time previous.

About 1857 Mr. Ralston, in conjunction with Garrisson, started a savings society, which ultimately developed into the Bank of California, with W. C. Ralston as president. He was also interested in the development of many mines and in the building of silk factories, sugar refineries and woolen mills, railroads, and steamship lines to Australia and China. It was in 1867 that Lawrence Barrett first met Ralston, whom he succeeded in getting to build a theater for himself and John McCullough.

The California Theater was accordingly built and opened in 1869 and for a while paid, but after a few seasons Barrett retired and the house was far from being a profitable investment. About this time luck seemed to be against Ralston. He had put money in the Palace and Grand Hotels, both of which were decided failures.

In August, 1875, James C. Flood made a sudden demand on the bank for $6,000,-000, and although they had assets amounting to a much greater sum than the one demanded, they were unable to meet the unexpected call. The bank was immediately closed and Ralston's resignation was requested. On the morning of August 27th he went out to North beach to take his customary swim and was never afterward seen alive. The general opinion was that he felt his disgrace so keenly that he committed suicide. He was one of the foremost men in the development of mining enterprises and was largely interested in the Comstock and other mines.

PIERSON B. READING.

The Man Who First Discovered Gold in Northern California.

To Major Pierson B. Reading belongs the credit of having done for Northern California what Sutter and Marshall did for the central portion of the State.

Born in New Jersey in 1819, he came to the coast in 1843, reaching Sutter's fort the following year, when he entered into the service of the General. In 1846 he obtained a grant of the upper portion of the Sacramento valley, which is included within the boundaries of what is now known as Shasta county. Here he settled and explored the wild country lying between that section and the ocean.

In 1848 he heard of the finding of gold at Coloma and hastened thither. After a short stay he became convinced that gold was to be found in the region which he had previously explored west of the upper Sacramento valley. He returned at once, organized an expedition, and soon found gold on the headwaters of the Trinity, thus opening a region which has continued to produce a large amount of the precious metal without interruption down to the present time.

Mr. Reading was one of the most prominent men in the State, and up to the time of his death none stood higher in the public estimation.

DAVID D. COLTON.

A Successful Manager of Many Large Mining Operations.

The subject of this biographical sketch was born at Monson, Me., July 17, 1832, and died at San Francisco October 9, 1878. As soon as he had attained sufficient age and preparation he entered Knox College, Galesburg. The innate individuality of his character asserted itself in his early youth, for we find him a successful teacher and fairly out in the world for himself at 16. At 18 he had started in company with a friend for California, where he landed in 1850. Soon after his arrival he commenced mining on the Feather river. He was stricken with typhoid fever and soon came down to San Francisco, and thence, in the latter part of the year, went to Oregon.

In 1851 gold was reported to have been found in Shasta butte, now Siskiyou county, and he, with hundreds of others, hastened there to prospect and locate mines. About this time one Charles McDermott became Sheriff of the county and appointed young Colton as his deputy. McDermott soon found more lucrative enterprises and turned over his office to the young under sheriff. By this office friends became numerous and money poured into Colton's hands by thousands. In 1858 he returned to New York with his family and entered the law school there, from which he graduated in due time. He then returned to California with a fine law library, accompanied by Ralph C. Harrison as his partner. Without being a politician he took a patriotic interest in public affairs. In 1865 he went abroad with his family, spending two years in traveling over Europe, the Holy Land, Turkey and Egypt. Mr. Colton was a large owner in, and president of the Amador gold mine and also of the Rocky Mountain Coal

David D. Colton.

Company. The Amador mine produced it's half a million dollars in gold bars annually and under his management kept up fully its reputation as one of the best managed mines on the Pacific coast.

He was also interested in many other mining ventures and was always a believer in the mines of this coast as a permanent source of wealth.

ANDREW J. BRYANT.

A Practical Miner Who Rose to a High Station in Life.

This gentleman led a long, successful and honorable life among the business men of California. He was born in Carroll county, N. H., in 1831, and worked on his father's farm until 1849, when he started for California, where he arrived in 1850. Upon arriving he went directly to the northern mines, where he remained about a year, when he was taken sick, which compelled his removal to San Francisco in order to obtain medical treatment. With health improved he engaged in the express business, which he followed for about two years, after which he removed to Sacramento and engaged in merchandising for a period of eight years. In 1866 he was appointed Naval Officer by Andrew Johnson and held the position until 1870. Mr. Bryant was a strong advocate of all measures for improving the condition of men who labor for a living. He was in favor of the eight-hour law, believing that eight hours was as long as any man should labor at hard work. The following is an extract from the *Shop and Senate* of August 23, 1875, after he had received the nomination for Mayor of this city: "He will be elected and he will make the best Mayor San Francisco has had since the time of E. W. Burr. He will not allow the city to be swindled either by gas or water companies. He will not encourage idle policemen or sinecure employes. He will break up that devilish den of Chinese that now corrupt and poison the very heart of San Francisco."

The majority of these things Mr. Bryant did accomplish with the exception of the abatement of the Chinese nuisance, and the power to do that lies not within the province of any one man, no matter what his position may be. San Francisco had an honest, devoted and popular magistrate in the person of Mr. Bryant. He presided over the deliberations of the

Andrew J. Bryant.

Common Council with dignity and impartiality, and had the moral courage to interpose his veto when he considered that that body was going beyond its legitimate powers. As a man he was generous of heart and courteous of bearing; as a citizen he was spirited and took a deep interest in any movement tending to better or elevate the community, while from the time of his own personal experience in the mines he took an active interest in the development of the mineral resources of his adopted State.

W. S. O'BRIEN.

One of California's Most Successful Pioneer Miners.

William S. O'Brien first saw the light of day in the picturesque village of Stradbally, county Dublin, Ireland, in 1826. When but a boy he bid farewell to his native land and emigrated to New York, where in 1847 he became an American citizen. After the discovery of gold in California he made a voyage around Cape Horn in the ship Tarolinta and arrived in San Francisco July 6, 1849. After working in the mines until 1851 he went into business with Colonel William C. Hoff, with whom he remained for about two years, subsequently entering into the ship chandlery business with J. W. Romer. In 1856

Mr. O'Brien, in conjunction with his former mining partner, J. C. Flood, opened a saloon on Washington street, near Sansome, which in after years became a favorite resort for mining men and stock dealers, and thus, in a quiet way, no doubt, the two partners obtained information from their patrons which was of great advantage to them. In the early days of San Francisco Mr. O'Brien was connected with the Volunteer Fire Department, and at one time was foreman of California Engine, No. 4.

His genial disposition and that uprightness which characterized all his dealings in business affairs won him the respect and admiration of his comrades, and, indeed, all those who know him. For many

William S. O'Brien.

years the firm of Flood & Brien invested in stocks in a modest way and succeeded so well that in 1867 they invested largely and made considerable money. They soon after sold their business in the city and engaged exclusively in mining operations. In 1868 Flood & O'Brien entered into partnership with John W. Mackay and James G. Fair. Prosperity attended them from the outset. Their Comstock claims were obtained at a cost of about $65,000, and it soon became evident that a magnificent property had been secured. During the great excitement in Bonanza stocks in 1875 the members of the firm made their colossal fortunes and came to the front the most successful operators in the world, obtaining complete control of the Bonanza mines, which they retained until their death.

JAMES C. FLOOD.

The Celebrated Nevada Bank Owes Its Existence to Him.

America is the land of opportunity, and the Pacific coast pre-eminently the region of lucky venture and pecuniary rewards for those who use the chances for fortune which, while they are open to all, yet reward only those who have faith, courage and perseverance. There is no better illustration of this than in the case of James C. Flood, who came to San Francisco in the ship Elizabeth Ellen in 1849. He was born in New York city on the 25th of October, 1826. He received a common school education fitting him for any ordinary business or commercial pursuit.

Attracted by the golden reports from the Pacific coast he left his home and friends and came to California. He had little money, but made up for this shortness in a plentiful supply of health, strength and a determination to succeed. He commenced mining on the Yuba river with a "rocker" with but moderate success. By practicing strict economy he managed to save $3000, with which he returned to New York with the intention of opening a business there, but finding his capital was too small to suit his purpose, he returned to California, where money was more plentiful and business energy better rewarded. In 1854 he formed a partnership with William O'Brien and their business relations brought them in contact with prominent mining men and stockbrokers, from whom they received important in-

James C. Flood.

formation relative to mining interests and mines. Profiting by this, they made successful investments in the Comstock ledge, and in 1862 secured large interests in the Kentuck, Crown Point and Belcher mines, from which they received enormous profits.

Soon after this a copartnership was formed with John W. Mackay and J. G. Fair, and this was the foundation of the business associations which afterward became so famous for financial success and great wealth. The same determination to make every project successful with which Mr. Flood started his business career was always his leading characteristic as a financier, and the great Nevada Bank owes its origin and success to Mr. Flood's fidelity to sound business principles and correct financial laws. While in business matters Mr. Flood was very positive in his dealings with the world, in private life he was a kind and generous friend and liberal to all charitable objects, giving without ostentation large sums annually to various benevolent institutions.

JOHN M. BUFFINGTON.

The Originator of the Present System of Mining Accounts.

The formation of stock companies some years ago for the more thorough development of mining ventures became a necessity. A man thoroughly suited to the occasion was found in John M. Buffington, whose thorough proficiency as an accountant, enabled him to plan and develop the most perfect system which now prevails in the various mining companies. He was born in Somerset, Bristol county, Mass., on February 15, 1818. His early education was received in the common schools, and at the age of 14 he entered the State Normal School of Rhode Island, where, after receiving a thorough course of English and mathematics, he graduated at the age of 17. He arrived in San Francisco June 13, 1849. For thirteen months he worked in the mines and was very successful, making as high as thirty ounces a day of the precious metal. With his earnings he started a large mercantile house in Stockton, which he was subsequently induced to give up in order to accept the position of secretary of several large mining companies. Under his careful thought and prudent experiment, the crude and imperfect style of accounts in mining secretaryship has grown into a system of completeness and perfection. He saw new and unknown mines start into prominence and once prosperous claims dwindle into obscurity. But in all these scenes of struggle and care, in all these conflicts of financial ruin, he stood beyond reproach and enjoyed to the utmost the confidence of all who knew him.

JOHN CONLY.

One of Plumas County's Leading Mining Men.

Self-made men have the same right to be proud of that distinction that they have to be proud of their good name, their business integrity or their personal honor. Such members of the community are usually men of latent ability and sagacity who, without the advantages of the favored few, carve out their own future. Such a man was John Conly. Energetic, striving and industrious he commenced life at the foot of the hill, and through honesty and perseverance climbed upward until he reached the top.

He was of Irish descent and first saw the light of day in the city of New York on March 7, 1826. While he was yet young his father died, leaving him the sole comfort and support of his widowed mother. At the age of 16 he left New York for California, where he arrived in August, 1849. After his arrival he spent two years at Mokelumne Hill, mining with average success. He then went to Hansonville, Yuba county, remaining there about the same length of time. In 1853 he settled in Plumas county, where he became extensively interested from that time on. He did more to develop the mines of that region than any other one man.

WILLIAM SHARON.

A Self-Made Man Who Achieved Wealth and Prominence.

William Sharon, whose name will be remembered so long as the history of the gold and silver mines of this coast lasts, was born in Smithfield, O., January 9, 1821. As a boy, he was a leader among his mates, but his individuality was too pronounced to make him popular. Although fond of play, he was also fond of his book, and found time to develop himself mentally as well as physically. He went to college, where he paid his way by doing odd jobs around the institution. He applied himself closely to his studies, was a ready debater and fully able to cope with his fellow students. But circumstances prevented him from completing his college course. After leaving college he returned to farming, which he soon gave up for the study of law, entering the office of Edwin M. Stanton, afterward Secretary of War under President Lincoln. He left his home and started for the Golden West April 1st, 1849, and arrived in California during the latter part of July of that year.

When silver was discovered in Nevada

William Sharon.

he was one of the first to aid in its development. At first the mines were a failure, but afterward Sharon and all connected with them amassed great wealth. When the Bank of California was established he became one of the trustees, and during the troubles of that institution, which arose through the death of its president, Mr. Ralston, Mr. Sharon succeeded in bringing the affairs of that institution to a satisfactory conclusion. He was elected to the United States Senate from Nevada in 1875 and held his seat until 1881. His Senate record was not particularly striking, as he made few speeches, gaining his points more by his actions than his oratorical powers. He had large property interests in San Francisco and owned the Palace Hotel, where he died November 13, 1885.

SAMUEL BRANNAN.

A Man Who Made Money and Died in Need of It.

Few names among the early pioneers of California have been more intimately associated with the history and development of the State than that of Sam Brannan. A review of the principal enterprises which mark the improvement and onward march of California would reveal him as their zealous advocate and promoter. Mr. Brannan was born in the town of Saco, in the State of Maine, in 1819. He immigrated to Lake county in 1833, where he learned the printing business, and in 1837 traveled through the State as a journeyman printer.

Five years later he published in New York city a weekly newspaper called the New York *Messenger*. As early as 1846 he formed a company to settle on the then almost unknown shores of California, and the ship Brooklyn, in which, with 230 immigrants, he sailed from New York, arrived at San Francisco in July of the same year. He at once became a leading and influential member of the isolated little community, and soon after his arrival he erected the machinery of two flour mills in a locality answering to that which is now known as Clay street. These were the first introduced into the country. In January, 1847, he published a weekly newspaper called the *California Star*, which was the first journal published in San Francisco, and was the parent of the now defunct *Alta California*.

In 1851 he was chosen President of the

Samuel Brannan.

famous Vigilance Committee, and in 1853 he was elected State Senator. He was one of the founders of the first school in San Francisco, and contributed liberally to the building of that edifice. In the silver mining districts of Eastern Nevada Mr. Brannan's business talents were also exerted. At Robinson's District he erected sawmills, quartzmills and smelting works, he built toll-roads and developed one of the richest mineral districts of that State. Of late years he did not cut a very prominent figure in public life, and for a few years previous to his death was seldom heard from, having removed to Mexico, where he went to look after his large landed interests.

His latter life was not as fruitful as his early years predicted. Wild speculations caused financial reverses, and he was left at the end of his days a poor man. He died in abject poverty in Southern California on May 6, 1889.

HENRY M. NEWHALL.

Another Pioneer Miner Who Achieved Success.

The career of this gentleman furnishes an apt illustration of the vicissitudes through which California has led many of her adopted sons to the prominence and comfort of their position in after life. He gained wealth and distinction, but only through the exercise of an energy and force of character which never deserted him. Henry M. Newhall, the fourth child of a family of eight, was born May 23, 1825, in Saugus, Mass. He attended school until he was 13 years of age, when he left home to seek his fortune. In the winter of 1849 he started for California, and after a tedious delay of several months on the Isthmus of Panama he reached this city July 6, 1850. At first he went to mining, but finding that was not his forte he left the mines and returned to San Francisco, where he entered the auc-

Henry M. Newhall.

tioneering business, which proved a successful venture. He made money rapidly, and in 1865 he became a large owner in the San Francisco and San Jose Railroad, which at that time was in a sorry condition. In 1866 he was elected manager, and from the time he took charge of affairs the success of the road was assured.

He never lost his interest in mining, however, and was interested in the development of the petroleum deposits of the southern part of the State, as well as in other enterprises which have done much to add to the wealth of California.

GEORGE W. KIDD.

One of the Originators of the Yuba Canal Scheme.

Captain George W. Kidd was born in Wayne county, Ky., February 28, 1813. He was the third in a family of seven children. His parents were exceedingly poor, and relying on a sterile tract of land for support, the captain's childhood and youth was an epoch of unremitting and severe toil. Captain Kidd showed a disposition to accumulate early in life; as a boy he was always saving, and generally succeeded in scraping small amounts together. However, these scanty sums were not sufficient to keep him on the farm, and at the age of 18 he bade farewell to his home and parents forever. He soon arrived at Cincinnati, O., where he sold his horse and secured a position as cabin boy on a river steamer. After working for seven years in that position he went into the business of rafting on his own account. About 1841 he visited Texas and spent a year in exploring that country. In September, 1849, he arrived at Sacramento. Until 1854 he was engaged in trading, freighting, mining, etc. The South Yuba canal was commenced by him and others in the fall of 1854 and completed in the latter part of 1857. The great success of this undertaking placed Captain Kidd and the other inaugurators prominently before the public. From that time he made money fast. Captain Kidd was a representative of that class of California pioneers whose courage and energy made the State what it is to-day.

JOHN A. PAXTON.

Many Enterprises Owe Their Successful Outcome to Him.

Like many an early pioneer, John A. Paxton owes his financial success to that adjustment of affairs which awards prudence, energy and sagacity. He was born in Rockbridge county, State of Virginia,

John A. Paxton.

June 3, 1819. At an early age he was compelled to rely mainly upon himself for his support and advancement. After having received what is known as a common school education he left his home and started out to earn his own living. At the age of 20 he emigrated to Texas, where he was engaged in the mercantile business. In 1849 he started for and safely landed in California. Having decided upon his arrival in the Golden State to make it the field of his future business operations he commenced his career with that untiring energy which has since been his chief characteristic. In 1853, with Mark Birmingham, he engaged in the banking business in Marysville. The firm continued in operation until 1863, when Mr. Paxton withdrew, and having formed a partnership with W. B. Thornburgh established a banking house in Virginia City, Nev. He continued his financial connection with the establishment until 1866, when he retired from the firm and returned to San Francisco, where he continued to engage extensively in mining operations, as he had during all his life on the coast.

L. M. PEARLMAN.

One of the Pioneers of Virginia, Nev., and His Enviable Record.

Among the pioneer mining men and merchants of Virginia, Nev., none was more prominent than L. M. Pearlman. He may be said to have assisted at the birth of the great Comstock lode, for he first settled at Virginia City in July, 1859, coming from California. He it was who built the first brick house in Virginia City. As a merchant dealing in general mining supplies Mr. Pearlman made a record for honesty which never was questioned. His word was as good as his bond, and the only fault he ever found with himself was that he was too generous and confiding, and the result was the accumulation of a big lot of assets in the shape of book accounts and loans that remain uncollected to this day. In those days every merchant necessarily dealt in mining stocks "by the foot," as was the custom, and many deeds to and from Pearlman for "feet" in Sierra Nevada, Ophir, Chollar, Yellow Jacket and other Comstock locations are to be found among the records of Storey county. Like many others Mr. Pearlman experienced the vicissitudes of fickle fortune, and he concluded in 1863 that Virginia was a good camp to leave, and closing up his extensive business he tramped off to Silver City, Idaho, whose mineral wealth was then attracting general attention. There he met many of the early Comstockers, and his store was general headquarters hail fellow well met for all Pacific coasters. He remained in Idaho, off and on, until 1869, when once more he caught the removal fever, and attracted by the glowing accounts of the fabulous wealth of the Original Hidden Treasure and Eberhardt at White Pine, he pulled up his Idaho stakes and transferred himself and all his belongings to the tender mercies of the Treasure Hill pogonip. "That was the worst move I ever made," said Mr. Pearlman to a friend while recalling reminiscences, "for inside of three years I had lost all my money and had to start over again." Mr. Pearlman isn't the only Pacific coaster who can charge up total loss to White Pine. But he started again with true Nevada and California grit, and concluded to woo fortune where before he had been successful, and in 1874 he found himself once more in Virginia City, where, in his absence, the Comstock had been redeveloped and became richer than in its infancy. He was finally successful, devoting his attention specially to mining. In 1879 the spirit of unrest again possessed him, and he went over to Bodie, and was soon possessed of as many locations and prospects as any man in the camp. During the next ten years Mr. Pearlman gravitated from one mining district to the other, ever on the lookout for a big mine—something in the ne plus ultra line, which all mining men look for and so few ever find. The Tombstone excitement found him in Arizona, and from there he naturally and easily drifted into Old Mexico, and back over the line into New Mexico. Finally, in 1890, he found himself back in his old stamping-ground in Idaho, whose mineral wealth had also been materially developed in his absence. The big developments made in the now famous Delamar mine (recently sold in London for $2,000,000), attracted his attention, and it was not long before he had secured a bond on some adjacent properties that Mr. Pearlman believes will develop into second and third Delamars, the foundation and characteristics being substantially the same. These mines are in the old Owyhee mining district, near Tramps, a station on the Oregon Short Line Railroad, and are known as the Howe, Manhattan and Lepley groups, consisting of 125 acres each, all the claims being either patented or in process of patent. The work in the Howe and Manhattan has been prosecuted by a tunnel, now in over 1000 feet, and by a series of crosscuts, drifts and winzes. The ledge, which is now seventy-seven feet in width, is of the same general character as the Delamar. As depth is attained the value of the ore increases in gold. For several hundred feet down from the surface the percentage of value is 30 per cent gold and 70 silver. The Lepley is also being developed by means of a tunnel and presents the same general characteristics as the Howe-Manhattan. The assays of

L. M. Pearlman.

ore vary with the progress of development. They have run as high as $440 and as low as $4 per ton, but Mr. Pearlman is satisfied that the general average will be sufficiently large to insure the payment of healthy dividends. Active development is being prosecuted in both properties and will be continued until it reaches a point that shall insure the future beyond question or peradventure. Some months since Mr. Pearlman visited London and gave English capitalists an option on the properties, since which the developments have demonstrated so much additional value that he is entirely indifferent whether the option shall be accepted or not. The Delamar is a steady dividend payer, and its shares command a large premium in the London market. Other mines in the Owyhee district are large producers, and altogether it looks as if Mr. Pearlman is to be congratulated on the possession of valuable properties that will enable him to rest from his long labors. He is now at the Idaho front with engineers specially sent out from London to make a thorough examination of both properties, and telegrams received from Mr. Pearlman within a few days are most satisfactory.

COLONEL ISAAC TRUMBO.

An Energetic and Successful Mining Operator.

If all the business men of San Francisco and of California patterned their methods after the subject of this sketch, Colonel Isaac Trumbo, the result could not fail to be greatly beneficial to the interests of the coast. A stereotyped biography of him is unnecessary, as his name and his record are too well known. During the years he has engaged in business in California and Utah he has invariably occupied a most prominent position in the financial world, his master strokes of policy surprising and at the same time calling for the commendation of the community.

Although not 34 years of age, he has time and time again formulated enterprises and carried them to a successful completion which have baffled the tact and ingenuity of more mature years. After a residence in Salt Lake City and before reaching his majority he gained an experience in a business and a speculative way which has since proved of almost inestimable value to him. Twelve years ago he returned to California and engaged in mining in Placer county. By adopting a most systematic plan of conducting the business he made an unqualified success of it. Before retiring from

Isaac Trumbo.

mining as a business he had made a handsome addition to his fortune.

Determining to settle in San Francisco as a permanent place of abode he immediately interested himself in the financial and commercial interests of the city, and his influence soon began to be felt. His quick intuition and keen foresight have proved invaluable to his associates in business, many dangerous rocks being thereby steered away from which would have wrecked the enterprise.

Probably one of his greatest feats was the overthrow of the Dresbach-Rosenfeld wheat combination. Colonel Trumbo took the most prominent part in blocking this gigantic scheme, the history of the deal being one of the most important and interesting in the annals of finance in San Francisco. He is interested in the various electric light companies and in the American Cracker Company, besides other leading concerns. His hand is also seen in railroad business, he having been the leader of the syndicate promoting the Short Line Railroad from Salt Lake to Los Angeles.

In politics he is recognized as one of the leaders of the Republican party in California, and has lately been honored by being appointed as an alternate delegate to the national convention. Another deserved recognition of his ability and worth was his appointment on Governor Waterman's staff, which position it is needless to say was filled with credit to the service and to himself.

Colonel Trumbo is most unassuming in his demeanor, and his easy manner of reception make all feel at home. He is most charitable to the needy as his many quiet deeds of kindness will attest. It is not surprising therefore that he has hosts of friends and has the unqualified esteem of his fellow citizens. He is certainly deserving of both.

JAMES G. FAIR.

A Man to Whom California and Nevada Owe Much.

A biography of James G. Fair is not necessary to bring him before the public mind. During the past forty-three years he has been a resident of the Pacific coast, and prominently identified with its industries, improvements, and especially its mining interests. No operator has been more

James G. Fair.

successful in detecting and tracing up indications of ore, perceiving the qualities of new machinery and adapting it to its requirements. His scientific knowledge and energy were the means by which he acquired his enormous fortune.

He exercised a personal supervision over all the mines with which he was identified, and received daily reports from his several managers to whom he gave each night instructions for the following day. Mr. Fair's commercial standing is of the highest. He is very popular for his genial and social qualities among his friends, as well as admired and esteemed among his business associates. Mr. Fair's political career as United States Senator was entirely satisfactory to his constituents, but was merely an incident in his life, which is that of an eminently successful mining operator.

J. L. GOULD.

A Hydraulic Mining Expert and His Vast Interests.

One of the pioneers of the State who has made a careful study of hydraulic mining from every standpoint is the subject of this sketch, J. L. Gould. Coming to California from Maine early in the fifties he located in El Dorado county and engaged

J. L. Gould.

in mining. He stayed in this county some little time, meeting with varying success, but not being satisfied he moved to Placer county, where he has remained ever since. Mr. Gould is well known throughout the State as organizing an almost perfect system of hydraulic mining. It was largely through his efforts and study that the rich mines in the ancient river bed running through Dutch Flat have been made to yield such enormous dividends. When, in reality, the work was but fairly begun, and every indication pointed to immense yields in future, the debris question came up, and as a consequence the mines have been shut down ever since. Mr. Gould has not confined his operations exclusively to mining. At different times he has been engaged in lumbering, farming, the construction of reservoirs, tunnels, ditches, etc., in all of which he is successful. He lives with his family in an elegant home in Dutch Flat, and while the place is at present practically deserted, Mr. Gould has pinned his faith to the locality, believing that with the resumption of hydraulic mining Dutch Flat will be one of the most thriving interior towns of the State. He is a most unassuming man, the doer of many quiet deeds of charity, and is universally admired and respected by the public at large and endeared to the hearts of his friends and acquaintances.

J. E. DOOLITTLE.

A Born Miner and a Miner All of His Life.

The subject of this sketch was born at the old Doolittle mine on the South Yuba river, one mile below the town of Washington, Nevada county, on January 10,

1856. His father was a pioneer of 1849, being one of the first arrivals during that memorable period. He settled in Nevada

J. E. Doolittle.

county and engaged in mining, and also built numerous roads, ditches, etc., besides making many improvements in milling and mining machinery. He was a thorough miner from start to finish, and as a consequence made mining a success. It is not surprising therefore that his son, closely following in the footsteps of his father, should also have made an unqualified success of the business. He has been engaged in mining the greater portion of his life, having mined in several counties in California, but settling in Dutch Flat as a permanent place for his operations. With his thorough knowledge of hydraulic mining he has been more than successful and had not the injunction been placed on this character of mining he no doubt would have built up an immense fortune. Mr. Doolittle is one of the most convincing talkers in the State and the strongest opponent of hydraulic mining is easily won over to his way of thinking after listening a short time to his arguments. He is a genial man, liberal to a fault, and has hosts of friends throughout the State. He accepts the present situation philosophically, believing that but a short time will pass when the miners will be given their dues and be allowed to resume work.

CHARLES S. WIELAND.

President of One of Amador County's Great Mining Companies.

In a review of the mining districts of California and the Pacific coast the same would be incomplete without notice being given to one of our most progressive young citizens, Charles S. Wieland. He is the son of John Wieland, the well-known brewer, and though but 25 years of age he has made himself one of the most representative business men in San Francisco.

On leaving college he became officially connected with the Philadelphia Brewery, where he first evidenced his business tact and ability. When the John Wieland Brewing Company was formed he was elected secretary, which position he held until the business passed into the hands of the present owners. As president of the Clinton Consolidated Mining Company his influence has been strongly felt in mining circles. This company owns some of the most valuable mining property in the United States, and, under Mr. Wieland's able supervision, the affairs of the company are prospering. The property of this company comprises what is known as the Clinton, Macato and the Clinton Peak mines. These mines are situated six and one-half miles east of Jackson, in the Pine Grove mining district. The claim has 4200 feet of ledges in it—one of 1200 feet and two of 1600 feet each. There are three large quartz veins, from fourteen to twenty-two feet wide, with cross-veins of from four to five feet in width.

The property of the consolidation consists of six full claims, each 1500x600 feet, and has been opened in places by both tunnels and shafts. The surface is worked by a gravity tramway, 800 feet in length to the mine. The two upper and lower beds upon the property are known as the Union and Paugh veins. The veins contain mostly ribbon rock, heavily charged with auriferous sulphurets, also galena, the same being estimated as high as $600 to the ton. A fine thirty-stamp mill is in operation at the mine. The mill is supplied with both water and steam power. The water power consists of a six-foot Pelton wheel working under a head of 166 feet. The steam power consists of a 12x24 Meyer's cut-off engine and a 64x16 horizontal tubular steam boiler. The water is conveyed to the mill by a pipe running 2000 feet in length from the ditch system. The stamps weigh 1000 pounds each and are of the finest quality made. Twelve Frue concentrators are also used to save the sulphurets. The ore bin has a capacity of 240 tons, and two black rock breakers, 9x15, are used to press the ore for the stamps. Some time ago the company decided to erect chlorination works in order to handle their own sulphurets. The works are now running with a capacity of six tons a day. Throughout the entire mine the same appearance of solidity is noted and everything necessary to the successful operation of the mine is looked out for. The company has lately added an Ingersoll-Sergeant ore compressor plant to run three drills, consisting of one 12½x14 Ingersoll-Sergeant piston in the cold ore compressor class "B," driven with water-jacketed cylinder and heads, and three Ingersoll clamp drills, with columns, clamps and accessories.

The town of Wieland, named after the

president, of the company, has lately sprung up, and from present indications a large settlement will soon be in existence there. There are 225 houses there already, with the prospects of a large number being built in the near future. A postoffice will shortly be established there, and it is the intention of the company to erect a schoolhouse and employ a teacher at their own expense in order to give the families of the mining element an opportunity to educate their children free of charge. In round numbers the company have expended about $300,000 in improvements to this property. It may readily be conceived that the past or present return from the mines or the certainty of future returns must be necessary to insure such a large outlay.

The ore averages about $6 per ton, most of it being at this rate. It takes second rank among the gold producers in this county, but one mine running ahead of this amount.

Mr. Wieland, besides taking a prominent place in mining circles, occupies an enviable position in social and fraternal relations. It will be remembered that during the vote for the most popular Native Son Mr. Wieland stood quietly in the background, never making an effort to gain a vote for himself. Notwithstanding this fact he was voted to be the most popular in the order, his majority running over 10,000. He is a member of Alcalde Parlor, No. 154, and this recommendation was highly appreciated by all of the members of this parlor.

Charles S. Wieland.

JOHN NOYES.

A Pioneer of the Butte and Cœur d'Alene Districts.

One of the most prominent men of the Pacific coast, who has long been identified with the early history of mining, is the subject of this sketch, John Noyes. Arriving in California in 1852 from Canada, he immediately started mining near Bridgeport, Nevada county. The method then in use was a pick and shovel, hand and rocker. He mined there about two months, but this method proving too slow for him, and the profit not suiting his ambition, he went to Nevada City, where he stayed for one and one-half years. He met with varying success in Nevada county, sometimes believing that a fortune awaited him in time, and possibly on the morrow looking face to face with poverty. Arriving in Virginia City, Nev., he prospected the country thoroughly, and stayed there until 1861. He was there before there was a single house built, and only a place which was seemingly a barren waste. Hearing of the immense richness of the mines in Idaho, he disposed of his interests in Virginia City and went to Florence. He immediately engaged in mining, staying there until 1865, at which time he went to Butte City, Montana, and for twenty-seven years he has made his home in that place, spending most of his time and energies in placer mining. While it is not known by the majority of people not directly interested in mining that Butte City ever offered facilities for placer mining, Mr. Noyes states that this was the principal industry at that time. The present site of Butte City has been mined by different companies, the present location of the gas works being worked by Mr. Noyes. Mr. Noyes has been very successful in these operations and has amassed a large fortune. Some time since he became interested in the Cœur d'Alene district, and as a result the Cœur d'Alene Silver and Lead Mining Company was incorporated. When this mine was in full operation it

John Noyes.

employed 140 men, the average output being about 250 tons daily. With the other mines affected by the strikes in this district, it is closed down, and about 2000 men are out of employment in this district caused from the strike.

Mr. Noyes is very hopeful for future results, both in Montana and Idaho, and says that he considers it the greatest mining region in the world. Mr. Noyes is heavily interested in real estate, both in Butte and in Seattle, Wash. In the latter place he is the proprietor of the Grand

Hotel on Front street, considered to be one of the finest hotels in the Northwest. Mr. Noyes has two sons aged 18 and 19 respectively, who are going to school in San Francisco, and upon completion of their education will, no doubt, follow in their father's footsteps and relieve him some from many business cares incident upon the care of his diversified interests.

JAMES F. TICHENOR.

Some of the Mining Operations in Which He Has Participated.

New York has been a prolific field for mining operations, as James B. Haggin and other well-known operators can attest. In 1882 there went to New York from San Francisco as a mine operator and promoter James F. Tichenor, who has since resided there, with occasional returns to the Pacific coast to visit mining properties in which he is interested. Mr. Tichenor was no stranger to New York, for he came originally from Jersey, just across the Hudson, and he had lots of friends in the great commercial center. The Tichenors are an old Jersey family whose history and traditions go back several hundred years, and whose record has always been a guarantee of honesty and good faith.

Mr. James F. Tichenor first visited San Francisco in 1870, when the mining excitement on California street was at fever heat. He speedily made the acquaintance of the leading operators and capitalists, among others General George S. Dodge, who was largely interested in the Eureka Consolidated Mining Company's operations. It was what may be called a very big deal, and Tichenor speedily secured a block of stock, which he subsequently sold at a large profit. His operations in Eureka naturally directed his attention to other Nevada mining districts, and he became largely interested in the Tuscarora district, and built the first stamp mill in that excitable mining camp. It was then that he learned his first object lesson in mining, for after the mill was completed and ready to run he discovered that he hadn't a mine to supply the ore. Many an older miner has made the same error, over and over again. The owners of the Grand Prize had a big mine but no mill, so negotiations speedily resulted in the mill becoming the property of the mine. Tichenor secured big blocks of Grand Prize stock, and when it was selling freely at $20 per share, realized sufficient

J. F. Tichenor.

not only to pay for the mill several times over, but a small fortune besides.

Mr. Tichenor's next successful venture was in the big Sierra Nevada deal which convulsed the San Francisco stock market in 1878. Probably more money was made and lost on the Sierra Nevada-Union deal than upon any operation connected with the Comstock. And here is where Mr. Tichenor learned a second lesson which has been taught over and over again to mine operators: "Clean up your stock then stay out." He obtained early information as to the situation of affairs, secured lots of stock, which he sold not exactly at top prices, but he realized a profit to the good of $150,000. Then came the temptation and the lesson. The break came, and from top prices the stock looked cheap and Tichenor went in again and lost some of his previous profits, but not all. "All things considered, Sierra-Union treated me fairly well," is Mr. Tichenor's general reflection.

In 1882 he went back to New York and wasn't long before he was floating blocks of Tuscarora stocks on the New York market. He made new friends rapidly, because advising them to buy Navajo at less than a dollar per share they realized it from $5 to $10. The Navajo deal, brilliant as it was, was followed by the North Belle Isle, Commonwealth, Belle Isle and other brilliant operations, in

which big profits were made on the Tichenor tips. It is only fair to state that in all his New York speculation Mr. Tichenor has been very successful, and his success has not been at the expense of his friends, leaving bitter reminiscences. They are always ready to invest when Tichenor gives the word.

The chief object of interest on the Pacific coast now for Tichenor is the development of a group of miles in Mono county, Cal., not many miles from the Bodie camp as the crow flies. Associated with him in this work of unearthing what they believe is a coming "winner" are John F. Cassell and Steve Roberts, who are pretty well known in San Francisco's mining circles. George Crocker is interested in the same property. From his confidential advices Mr. Tichenor is satisfied that the new property will pay handsome dividends when the work of development shall be completed, but there is no disposition to market the stock at the present time.

It was through Mr. Tichenor's earnest efforts that Port Orford, Or., was declared by the Government to be a harbor of refuge. His uncle, Captain Tichenor, possesses large land and lumber interests at Port Orford, in which his nephew is also interested.

JAMES R. KEENE.

A Man of Indomitable Perseverance and Sterling Integrity.

One of the most picturesque figures in that remarkable group of speculators, financiers and adventurous spirits whom the riches of the Comstock drew to San Francisco from all parts of the world, in 1860 and 1861, was James R. Keene.

Crossing the plains in '52 or '53, he arrived in California at the age of 13, toughened by the exposure involved in that long journey. He brought with him a fair knowledge of Latin and French, and a thorough knowledge of his own language, acquired through years of drilling in the English schools. His father, who was a merchant in London, had suffered business reverses, and brought his family to the great El Dorado to start life anew.

There was nothing in the traits or appearance of this boy to suggest that he was destined to become one of the foremost speculators not only in California, but in the East, or that he would be known among leading financiers the world over.

His first employment was with the Government at Fort Redding, where with many others he engaged as cowboy to herd Government mules and cattle in the Indian country. Tiring of this, after a month or two, he caught the fever which attacked every one in those days and determined to try his chances at placer mining. This he followed with indifferent success for several months, returning home to find that fire had literally swept away the town where his family had settled, and with the home had gone all the family's possessions as well, and the population of the destroyed hamlet had been obliged to camp about the hills in tents. Securing a team, the lad joined in the general work of hauling lumber and such other materials as were necessary to rebuild the town. When this work was over young Keene went into the cattle raising and dairy business. Then later he bought a sawmill and a flouring-mill and, in fact, turned his hand to everything in his ambitious desire to help along the fortunes of his family.

Making no money worth speaking of in these various undertakings, he tried the newspaper business and for a couple of years edited a local paper rather successfully, but this brought little fame and less revenue. Meanwhile he occupied every spare moment in general reading and in the study of the law, and before he had ceased to be an editor he had strengthened his mind with an acquaintance with every book which came within his reach.

Just at this period came news of the great discoveries upon the Comstock lode, and Keene started for Nevada. There, for the first time, he found ample field for his activities. He went to San Francisco, and in the course of a few months made his first $100,000, returning home to marry Sarah Dangerfield, sister of the late Judge William P. Dangerfield. Back to San Francisco, actively engaged in business, he lost all that he had made in the great collapse following the first inflation in Comstock shares, and found himself in debt for a large amount of money. For more than a year following he had a hard struggle. All his friends had lost their fortunes, too, in the general wreck, and he became acquainted with sorrow in its severest forms. But his dauntless spirit was not overcome. He refused all offers to engage in business, keeping his eye on the stock market, feeling that in that field was his only chance, measured by the estimate he had of his own speculative capacity. He finally succeeded in obtaining a position with William Burling, who was the great broker at that time; and to young Keene Burling gave the execution of all of his vast business outside of the Stock Exchange. He showed such ability, fidelity and secrecy in the execution of these orders that he was offered by Charles N. Felton, one of the present United States Senators from California, his seat in the Stock Exchange, with the understanding that whenever Felton called for the price of his seat Keene should pay for it, a kindness that until this day Keene remembers, and the two men are still close friends. Keene had already made considerable reputation as a broker and operator, and the advantage of a seat in the Stock Exchange gave him opportunities to establish himself in a short time as one of the most active and prominent members of that organization.

He paid his debts. idly. In twelve m[onths] into the Stock Ex[change] partnership made w[ith] of San Francisco, di[ed worth] $400,000.

The struggle and [he] gone through told upon his constitution, which was never strong, [a]nd his nervous system broke down, obliging him to quit active business for some months to recruit his health. When he returned he went into speculation with his old dash and enthusiasm, literally lifting the stock market, which he found absolutely inert and dead, into a state of great activity; and in the few months following he added largely to his accumulations. Again by the order of his physician he was compelled to make a sea voyage to China. He left behind a large amount of the stock of the Consolidated Virginia, which he had become convinced was destined to be a largely productive property, and consequently advance in value. Not only in this was he not mistaken, but when he returned the shares rapidly advanced until they reached the phenomenal figures so well known to all dwellers on the coast and participants in that wonderful speculation. Selling out his stock on the top of the wave Keene found himself with a fortune of nearly $6,000,000, but with shattered health and a realization that for a time at least he must abandon his active career. The average operator would have been well satisfied with the results.

During this period an event occurred which was fraught with the greatest consequences to the financial and commercial interests of the Pacific coast—the failure of the Bank of California. Having no personal interests whatever in the institution, Keene recognized immediately the emergency which confronted the commercial and mining communities, promptly subscribed $1,000,000 personally, rallied his friends in the Stock Exchange and passed a resolution, against most formidable opposition, that the Stock Exchange should contribute half a million more, which it did, and then by personal solicitation secured from brokers and capitalists a further sum of three-quarters of a million to the relief syndicate of $8,000,000 organized to rehabilitate the bank. The bank reopened its doors, business resumed its normal state and California was rescued from what threatened to be a great calamity. To the members of this syndicate the people of California owe a lasting debt of gratitude.

Exhausted by the anxieties and excitements of his large speculations, Keene determined to go abroad, and left California in 1877. Finding his health somewhat improved by the change when he reached New York, he determined to remain a few months before crossing the Atlantic. Naturally his restless spirit led him into Wall street, and he was soon speculating with the same freedom that had characterized his stupendous San

least. One company of these people is known to have taken two years ago from their claim on the Klamath below the mouth of the Scott river, over $100,000. As a rule, however, nothing definite can be learned in regard to their earnings, which are probably much larger than is generally supposed.

There are instances in which the production by some of the white companies [at this] time was too great a temptation for Keene to resist, and he forthwith purchased an immense line of the better properties, displaying a knowledge of their selection on the short acquaintance he had with them which no one could have acquired without possessing that inspiration which was born in this remarkable man.

In fifteen or sixteen months Keene accumulated a profit of nearly $10,000,000 and sold out his stocks and pocketed the money. Again he resolved to abstain from business. But he had interests left in Chicago through an entanglement with other persons in a speculation in wheat. Flushed with his tremendous California and New York successes, strong in the possession of so much available money and indignant at what he deemed a piece of treachery on the part of one of his coadjutors, he in a moment of weakness permitted his temper to get the better of his judgment and joined a syndicate to control the wheat of the world. The magnitude of the operations of this combination became so conspicuous that it riveted the attention of the country and brought down upon the undertaking the anathemas of the entire press.

Purchasing in all the interior markets, the whole stock of wheat of the United States (outside of California) was under its control, and every one remembers how England and all the other European countries sat down and refused to buy.

The newspapers took up the warfare actively against the combination. Keene's allies disregarded their contracts and agreements and did not hesitate to sell out their holdings, but secretly; and he, with all the world arrayed against him, stood up and faced the consequences at a cost to him of $3,000,000 in less than sixty days.

He was still a rich man notwithstanding this dreadful drain. But his prestige was dimmed, and no one knew better than he what this meant in Wall street. Yet it is related that people who called upon him at this most critical period in this eventful drama found him always perfectly unruffled, although the wildest stories attacking his solvency were being spread wherever telegraph wires would carry the "news." But under the calm exterior—which no man who ever lived could maintain with more composure—the iron had entered his soul. From that time he engaged in operations in stock privileges on a scale so great that even his best friends felt that the time would come sooner or later when disaster would surely overtake him. This came in 1884, when he was compelled to suspend payment, producing with other failures at the same time a tremendous

In an operation on the market, which was [ma]de in the course of a [deal] credited with making prices of all railroad

panic and .... him with ....
aggregating between $1,500,000 and $......,
000.

How this man has succeeded in wiping
out this vast indebtedness and again enrolling himself in the list of millionaires
is something which causes Wall street
never-ceasing wonderment. But, great as
is its admiration for the industry and
genius which have produced these results, it is less than the regard which it
has for Keene's absolute integrity and the
respect it feels for the fortitude and
courage which through years of adversity
sustained him and marked his conduct
and his bearing.

In reviewing a career so exciting and so
full of dramatic interest it would be of
value if a writer could accurately point
out its controlling characteristics.
Whether it is genius or the result of calculation, industry and the boldness to put
in execution at the proper moment the
plans formed in such a fertile mind it is
difficult to determine. There will be a
difference of opinion always on these
points, as there will be regarding the wisdom of some of the undertakings which
brought disaster to his fortunes. But
aside from these considerations James R.
Keene will always be regarded by those
best capable of forming an opinion as a
leader in the great movements in the
stock markets who never had a superior;
and despite the rivalries and antagonisms
necessarily involved in such an active
career there will be few found to dispute
that he must always be assigned a place
in the first rank of that conspicuous
coterie of men who in Europe and America
have at various times controlled the
course of prices in the exchanges of the
world.

E. M. HALL.

A Well-Known Mining Secretary and Broker.

For many years Edward M. Hall has
occupied a prominent position in the
mining community as secretary and as
the buyer and seller of mining stocks.
Probably no man on the coast is better
versed in the many intricacies of the
various stock deals which have taken
place in the San Francisco Stock and Exchange Board. He has by close application and careful study familiarized himself with the mining interests of the coast,
and is consequently in a position to talk
intelligently on the subject of hydraulic
mining. Mr. Hall is a strong advocate
for the resumption of hydraulic mining
and predicts that when the present difficulties are overcome that the business interests of California will be given a great
impetus. Like other thinking men, he
holds that the enormous amount of money
which could annually be taken from the
hydraulic mines could not fail to be of the
greatest benefit to the State at large in
the way of giving employment to thousands of men and creating a market for
manufactories and, in fact, almost every
class of business.

J. H. NEFF.

An Active and Intelligent Promoter of Mining Interests.

J. H. Neff, the president of the Miners'
Association, is one of the best-known men
in the State, and, while living in a small
interior town, Colfax, makes his influence
felt throughout the State. He is
of sterling qualities, his manage....
the Republican convention in S....

Jacob H. Neff.

while in the chair demonstrating h..
clearheadedness and keen foresight
Mr. Neff has been and is to-da..
one of the hardest workers i..
the cause of the hydraulic miners, and t..
his untiring efforts much of the success o..
the present agitation must be a..
cribed.

He is never discouraged, even when th..
lookout is the darkest, and, believing a..
he does that the miners cannot fail to b..
successful in their present endeavors, h..
invariably whoops up the courage an..
hopes of some more inclined to look f..
failure. "If by any chance we should no..
get relief from this present Congress,"
said Mr. Neff, "it is merely delaying i..
for a season, but it will come eventually.
Of course these delays are aggravating
and also expensive, but we have the consolation of knowing that we are so near
the goal that any day we will receive what
we have been working so long and earnestly for."

A. WALRATH.

He Is Heartily in Favor of a Resumption of Hydraulicking.

It must not be supposed that the advocates for a resumption of hydraulic mining are confined to the class directly interested in this feature of mining. On the
contrary, the quartz miners are to a man
in favor of it. In conversation with A.
Walrath he stated that while he was not
directly interested in hydraulic mining,
he thought the present agitation a good
one, and one in which he would give his
heartiest co-operation and support. Mr.
Walrath is a quartz miner, having recently disposed of his interest in the
noted Providence mine. While the shutdown of the hydraulic mines have not affected him personally, he is very decided
in his views on the subject, claiming that
the discontinuance of hydraulic mining
has been one of the greatest and most expensive mistakes the coast has ever ....

His views are certainly verified by the statistics with which he confronts you.

COLONEL A. ANDREWS.

One of San Francisco's Oldest and Best-Known Business Men.

The discovery of gold in California with the phenomenal strides to prosperity made by the State since that time form one of the most interesting as well as one of the most important chapters in modern history. The lives of the sturdy and venturesome pioneers occupy a conspicuous part in this history, and probably none stands out more strongly as synonymous with ability and integrity than the subject of this sketch, Colonel A. Andrews. This gentleman has since the date of his landing here on September 5, 1849, been closely identified with the interests of the coast, and his public spirit and generosity have long been a subject of most favorable comment. No popular movement whereby the State at large would prove the gainer has been started without his hearty cooperation and support, which has been so freely given. Whenever it has been found necessary to procure financial support from the public in aid of any movement Colonel Andrews has always been found to be one of the first on the list.

Colonel Andrews was born in London, England, in 1826, but while yet in childhood his parents emigrated to the United States and settled in New Orleans. Here he received his early education and was given his first lessons in business. How well he profited by them can be seen by his present standing in the commercial world. At the outbreak of the Mexican war, young Andrews, then a lad of 20 years, was appointed to a lieutenancy in the Second Ohio Regiment, but his adaptability to the profession soon made itself noticeable and he was shortly afterward promoted to the post of captain. His services during the war proved most valuable and his counsel was always eagerly sought. At the close of the war his worldly possessions consisted of a land warrant of 160 acres and the small sum of $250 in cash. His natural independence asserted itself, and instead of seeking employment he settled in St. Louis, and with the aid of his old friend, Michael Reese, opened a jewelry store. His reputation for conscientious work and honorable methods spread quickly and he was most successful in the undertaking. On hearing of the discovery of gold in California he at once determined to embark for the new El Dorado, and on September 5, 1849, reached San Francisco. Starting for Sacramento he established himself immediately in the jewelry business, and up to 1856 conducted the largest and most successful institution in that city. On October 8, 1856, he was honored by being appointed Colonel on Major-General Sutter's staff. This was the first staff organized in California, and Colonel Andrews was the first appointed on the staff. But four of this body are still living, namely, Colonel Andrews, Colonel Moulder, Colonel Richard Sinton and Colonel L. L. Warren.

Closing his jewelry business in 1856, Colonel Andrews spent eight years in traveling and in different speculations, embarking in many enterprises with varying success. In 1872 Colonel Andrews opened the celebrated Diamond Palace on Montgomery street, the fame of which institution has since spread all over the civilized world. The name adopted for the establishment by Colonel Andrews,

Model of the first nugget found at Coloma.

the Diamond Palace, is certainly most appropriate. On entering the doors a perfect blaze of light is reflected from the large number of diamonds and precious stones, set, unset, in cases, reflected from the mirrors, and the diamonds and precious stones composing a portion of the ornamentation of the exquisite oil paintings in the dome. It is truly a wonderful sight, and must be seen to be appreciated. Colonel Andrews has in the conducting of this establishment demonstrated his wonderful aptitude for business, as he has built up and maintained such a reputation for the Diamond Palace that it is now justly considered one of the most interesting sights in San Francisco, and it is no uncommon request to hear from strangers in the city to be shown the place, they having heard about it from fellow travelers.

Colonel Andrews has always made it a strict rule in his business to carry goods of the best quality and workmanship, and, having once gained the confidence of the public, it is not surprising that his business to-day ranks first in this line on the Pacific coast, and, for that matter, will challenge comparison with any like establishment in the world.

An interesting relic which Colonel Andrews has is the first piece of gold discovered in California. It was picked up by Captain John Marshall at Coloma in 1848 and presented by him to General S. who presented it

Colonel A. Andrews.

to Colonel Andrews. A fac-simile of the nugget is reproduced in the accompanying cut. Colonel Andrews has had many offers for this relic, but he has steadfastly refused to part with it. This small nugget may aptly be termed the starting point of California history. Colonel Andrews' generosity is well known, and his appreciation for bravery was shown when he presented to an engineer on the North Michigan and Lake Shore road a magnificent gold watch and chain as a testimonial of the bravery and presence of mind displayed during what looked like a terrible railroad fatality. Colonel Andrews never makes a demonstration of his charity and liberality, but takes it as a matter of course. The wisdom of the selection of Colonel Andrews by President Arthur as commissioner to the New Orleans Exposition in 1885 and also to London, England, in 1886, was shown by the splendid work done there, and the handling of California products by him has been of vast importance to the State by inducing immigration. A point well worth mentioning is that on his departure he was given a check for $10,000 to carry out the work. It is a well-known fact that California made the best showing at the exposition, and yet Colonel Andrews returned $3000 to the State. Colonel Andrews is an honored and respected Mason of forty years' standing, and has been connected with various civil and military organizations of this city, belonging in fact to thirty-three different organized bodies. On May 8, 1890, he was honored by being elected president of the Associated Veterans of the Mexican War, which position he now occupies. He was also the last president of Tammany, and first vice-president of the Manhattan Club.

At present he is president of the Eureka Building and Loan Association, a company with a capital of $3,000,000, and also president of the Northwest Gold and Silver Mining Company of Buck's Bar, B. C. He was elected to succeed Colonel Harney, who was killed in the railroad accident at Tehachapi. Taken altogether Colonel Andrews' record as a soldier, a business man, and as a citizen is without a blemish, and his present enviable standing is certainly deserved. He numbers his friends by the thousand, and is daily adding to the number, as it is said he never meets a man without making a friend of him. Since writing the above we see that Colonel Andrews has been elected a delegate to the national Democratic convention at Chicago, for which place he leaves in about two weeks.

HON. H. F. BARTINE.

A Stanch Congressional Champion of Silver.

No man in public life has proved himself an abler or more influential friend of the silver miners than Congressman H. F. Bartine of Nevada. He is now serving his second term in the House of Repre-

H. F. Bartine.

sentatives, and no member of Congress has made a better record so far as close attention to the interests of his constituents and the country at large is concerned.

His first notable work in Congress was his able and successful attack upon the Treasury ruling by which Mexican lead was admitted free of duty to this country. Such a proceeding would be, and was while it lasted, fatal to the lead mining interests of the United States. Those Americans whose mines were thus rendered valueless, made a strong protest, and found in Congressman Bartine an able champion. Free lead from the slave-worked mines of Mexico had its friends in Congress, and strong ones they were too. But the Nevada representative kept up the contest until he won it, and a duty of 1½ cents a pound was levied upon the Mexican product, thus saving to the United States one of its most important mining interests.

As a member of the Coinage Committee of the present Congress, Mr. Bartine has won a national reputation by his masterful struggle for the remonetization of silver. Particularly notable was the able manner in which he met the noted economist, Edward Atkinson, the champion of the gold bugs, who was obliged to retire from the field completely vanquished by the unanswerable arguments advanced by the member from Nevada. In many other ways has this gentleman won the respect and admiration of the friends of silver.

Congressman Bartine is emphatically a self-made man, and is a worthy representative of the West, which is his home. Born in New York City in 1848, he attended the common schools, and when but 15 years of age he enlisted as a private in the Eighth New Jersey, serving for two years in the Army of the Potomac, and receiving a severe wound at the battle of the Wilderness, but recovering in time to participate in the glories of Appomattox.

In 1869 he removed to Nevada, where he served a partial apprenticeship in mining as a mill hand, in the meantime devoting all his spare time to study. In this manner he acquired a good education, and in 1876 began to study law. In 1880 he was admitted to the bar, and was a candidate for Justice of the Supreme Court at the last election, only refusing that to accept a renomination to Congress, where his good work had endeared him to his constituents. He was elected to the Fifty-first Congress by a large majority, and again to the Fifty-second, receiving 6610 votes against 5736 cast for George W. Cassidy, Democrat.

J. B. HOBSON.
One of the Most Prominent of the Hydraulic Miners.

John B. Hobson, who has been engaged in California mines and mining in a practical and scientific way for many years, has just returned from Washington, where he did missionary work for the debris bill, the passage of which will be of such incalculable benefit to the State. If the measure become a law its success will be largely due to the logical, matter-of-fact way this gentleman presented the case to shy Congressmen. No better man for such a mission could have been selected by the California State Miners' Association. Mr. Hobson is 47 years old and a Dublin Irishman, which means that he is an educated gentleman. His father was that before him. In 1857 Mr. Hobson went to school in San Francisco, subsequently taking complete courses in mining engineering and chemistry under Thomas Price. From 1868 to 1872 he was a contractor on the State Capitol, and in the latter year he went to Placer county and engaged in mining and mining engineering, engaging in large hydraulic enterprises, as well as ledge and quartz mining at Iowa Hill. For two years he has been engaged as field geologist of the State Mining Bureau. He was one of the organizers of the hydraulic miners' movement in Placer county which led to the sending of a delegation to Washington, has always been a friend of the miner, and has been uniformly successful in life.

J. B. Hobson.

JAMES L. FLOOD.
Prominent in the World of Finance and Commerce.

James L. Flood, the subject of this sketch, is a native Californian, his father, James C. Flood, being well remembered as one of the bonanza kings associated with James G. Fair, William O'Brien and John W. Mackay. At an early age young Flood evinced unmistakable signs of following in his father's footsteps and being a fit representative for his parent's vast interests. It has so often been said that wealth to a young man is a curse that the public is inclined to believe this to be an accepted fact. It may be said, however, of a majority of the native sons who have been so fortunate as to be born of rich parents that they have done more to dispel this illusion than any other class of young men in the world. James L. Flood is certainly one of this class. Possessing a naturally keen business mind, which has been greatly sharpened by contact with the world, he has settled himself to the task of managing the immense interests left him in a manner that will always leave the balance on the proper side of the ledger. His success is too well known to need any very lengthy comment. Sufficient to say that no one could have acquitted himself more creditably, Mr. Flood is at present a director in the Nevada Bank, one of the strongest commercial institutions on the Pacific coast. During the term of his directorship he has demonstrated to the complete satisfaction of all concerned that his business tact and foresight are far above the average, and that the confidence reposed in him has certainly not been misplaced. His record as a financier places him in the front rank of the wealthy men of the State, and through all he has shown a spirit of fair dealing that redounds greatly to his credit. Notwithstanding his large interests Mr. Flood is thoroughly unassuming in his manners, every one feeling perfectly at home when in his company. He is the doer of many quiet charitable actions, notoriety in this regard being extremely distasteful to him. By his genial manners and his liberality he has endeared himself to the hearts of his associates, and by his business qualifications gained the confidence and esteem of the community. He is certainly a representative young Californian.

HENRY MARTIN.

A Warm Friend and Promoter of Mining Enterprises.

Henry Martin of Trinity county, while not a hydraulic miner, has shown a great interest in the movements of the State Miners' Association and by every means in his power given them his heartiest cooperation and support. Mr. Martin's mines are quartz mines and are situated in Trinity county, in which place the hydraulic mines are not affected by the injunction placed on this character of mining, as the tailings all run into the Pacific ocean.

Mr. Martin is very strong in his denunciation of the complete shut-down of hydraulic mining, and has given both his time and money in the endeavor to gain relief for those thereby affected. In conversation with a CHRONICLE reporter he said: "I consider that the injunctions preventing the hydraulic miners from working their property are nothing more nor less than a virtual confiscation of the property. It certainly is not justice for one class of men to be protected by the Government at the expense of another class, especially when the latter class is given no redress or reparation for the damage done them. The serving of this injunction on the hydraulic mines has been the means of throwing thousands of men out of employment; the number indirectly affected being much larger than the actual number of miners thrown out of employment. It has taken from $8,000,000 to $12,000,000 annually away from the circulating medium of California and has been an indirect injury to every merchant, business man, financier, and in fact every one whose interests are in California. Taking as an average $10,000,000 per year, which is the most conservative estimate yet made by engineers competent to speak on this subject, the grand total would amount to $100,000,000 which the State has been deprived of since the mines were shut down. This fact speaks for itself and is too plain a showing to need any comment. I firmly believe that with the resumption of hydraulic mining the business interests of the coast will be given an impetus which will result beneficially to almost every man who is interested here. The feasibility of the retaining dams has already been demonstrated, and it certainly is the duty of the Government, after depriving the miners of their dues for so many years, to help them out to this extent."

TIMOTHY GUY PHELPS.

Another Practical Miner Who Has Worked His Way Up.

One of the most popular of the pioneers of '49 is the subject of this brief sketch, Timothy Guy Phelps, United States Col-

Timothy Guy Phelps.

lector of Customs at this port. He is a New Yorker by birth, but being seized with the gold fever in 1849 he joined the venturesome spirits and came to this coast, and immediately engaged in mining. He spent some little time in the southern mines, but, his health becoming impaired, he returned to San Francisco and started in the mercantile business. In 1856 he was nominated and elected a representative of San Francisco and San Mateo counties for the Legislature. He was afterward elected to the State Legislature, during both of which terms he made an enviable record for himself.

In 1861 he was elected to Congress, and in 1869 was appointed Collector of Customs in San Francisco. Besides holding this position, he is a member of the Board of University Regents and of the California Pioneers, besides holding other valuable offices.

W. C. RALSTON.

An Active Promoter of Hydraulic and Other Enterprises.

On Thursday, November 25, 1886, appeared the first copy of a weekly paper called the *Mining and Industrial Advocate*. The paper was owned by W. C. Ralston and J. B. Hobson, and its name signifies its purposes. It is not generally known, but this issue was the starting point of the present agitation on hydraulic mining. Mr. Ralston had for some time prior to this been engaged in mining, having been connected with the Ralston Mining and Ditch Company, in Placer county. When the restrictions were placed upon the hydraulic mines he left the county and, coming down to San Francisco, gaged in journalism. Not having enough money to carry the undertaking through as he wished, he gave it up, and accepted the superintendency of a drift gravel mine in Placer county for a French company, which position he filled until appointed a notary by Governor Markham, at which time he came to San Francisco. Mr. Ralston is certainly entitled to the credit of starting the present movement by his work years ago. When the present call was made he responded at once, and, although his hydraulic mining interests were so small that it was not worth his while to spend any time on them, he did the major portion of the work in San Francisco preparing for the convention. His selection as secretary of the association was unanimous, and it is safe to say that no man on the coast stands higher with the mining element than W. C. Ralston. He is a hard worker and a persistent one, both of which qualities have proved of great value toward the successful organization of the association. Mr. Ralston is most unassuming, and really does not seem to think he has accomplished a great piece of work. The public know what his services are worth, and he is appreciated accordingly. His selection as United States Appraiser is another deserved recognition of his worth and ability, and he has the entire confidence of the public that his work will be well and faithfully done. He is a member of the Technical Society and the American Institute of Mining Engineers, New York. He is also a member of California Chapter, No. 5, Royal Arch Masons, and of the Stanford Parlor, Native Sons of the Golden West.

W. C. Ralston.

SIMEON WENBAN.

The Owner of One of the Best Silver Mines on the Coast.

One of the pioneer silver miners of the Pacific coast is the subject of this sketch, Simeon Wenban. His present fortune and success place him in the front ranks of the self-made men of the Pacific coast, for Mr. Wenban can fairly lay claim to having absolutely no assistance whatever from outside quarters in pushing his business to a successful completion. For years, he struggled against adverse fortune, and when a majority of other men would have long since given up the fight he continued his work until success at last crowned his efforts. Up to 1865 Mr. Wenban followed the business of mining without meeting with any great success, but since that period his mines have been among the best paying in the State of Nevada. He is the owner of several mines, the majority of them not now being operated on account of the cost of production and delivery and the low price of silver.

GARRISON MINE, NEVADA, PROPERTY OF S. WENBAN.

GARRISON MINE, NEVADA, PROPERTY OF S. WENBAN.

Some time since he made a general assay of the products of his mines, the average being $64 per ton. This may be said to be all silver, as the amount of gold carried in the ore is almost nominal. During the period of the full operation of his mines he gave employment to 150 men, the number of souls being indirectly supported by his enterprise numbering fully 500 and possibly more. No man in Nevada is more deservedly popular with his employes. It is true that enormous profits have been realized from the mines, the average yearly net profit amounting to $500,000. Mr. Wenban has always paid the top scale of wages, and, furthermore, always paid particular attention to the comfort and safety of his employes. At present, in the condition of silver mining in Nevada, he is employing but twelve men, or merely enough to look out for the property and watch his interests.

It was in 1862 that Mr. Wenban first made his start in Nevada. Proceeding to Virginia City he accepted the position of superintendent of a quartz mill. In 1863 he resolved to work for himself, and joined a prospecting party at Austin. After enduring may hardships the party arrived at Mount Tenabo, and located fifty-six claims. The Cortes joint stock company was at once organized, and in 1864 an eight-stamp mill was erected and the ores worked by the wet process, the same as on the Comstock. After $200,000 had been expended and but $20,000 taken out the works were closed down. Not disheartened he went to work by himself and secured a part ownership in four claims in the Limestone district, which claims he afterward purchased for $14,000. While on the road to fortune, a fire destroyed all his books, papers and title deeds to the claims. He then went to work, located the claims and obtained United States patents for them. As fast as money was taken from one mine he would put it into another and by so doing developed all his claims. Wenban finally secured, all told, 800 locations on easy terms and then purchased the mill of the Cortes company which had originally cost $100,000, for $10,000. It was used until 1885, when a new mill was erected. To gain an adequate water supply he bored two artesian wells, costing $30,000 each. In 1886 he substituted the amalgamating for the leaching process and found a great saving thereby. In all about $3,500,000 had been taken out of his mines up to 1888.

THE HIBERNIA BANK.

San Francisco's Great Savings and Loan Society.

The Hibernia Savings and Loan Society was organized in April, 1859. It was intended to be a local concern doing business with and for the people of this city and county and its immediate vicinity. This idea has been carried out, and with resources exceeded only by one institution of the kind in the United States it is a bank of which San Francisco is and ought to be very proud. All its money is loaned in this city, Oakland and Alameda, except about $2,000,000, lent on country property in this State. When it is remembered that the total amount of money secured by first mortgages is $20,531,276 it will be seen what a large monetary interest the bank has in San Francisco and the city in it. The total assets of the bank as published in the sworn statements of the Bank Commissioners for the year ending December 31, 1891, are $29,493,021.50. Of this large sum the bank holds in United States bonds and other bonds of acknowledged standing the sum of $7,230,870.91, the actual value and no the market value being taken into account, while of the class to which the bonds belong it is well known the mark value is considerably in excess of the actual value. In city real estate the bank owns its buildings on the corner of Montgomery and Post streets and on the corner of Jones and McAllister streets, the former valued at $200,000 and the latter at $543,503, very safe investments. At the date of the statement the cash in the vaults of the bank amounted to $507,637. In this estimate of the bank's assets the cautious, safe principles of the conduct of the society are illustrated, and if they were realized on to-morrow the assets would be much in excess of the handsome total quoted.

In a few weeks the bank will move into its handsome new quarters, one of the most substantial and, for its size, costliest buildings ever erected in this city. James R. Kelly is the president of the Hibernia Savings and Loan Society and Robert J. Tobin is the secretary.

JOHN C. QUINN.

The Achievements of One of California's Self-Made Men.

John C. Quinn, who occupies the position of Collector of Internal Revenue for this district, is a native Californian, hav-

John C. Quinn.

ing been born in El Dorado county in 1859. His chief education was acquired at the public schools in Nevada City, but by outside study and close application he has gained that scholarly attainment for which he is well known. When quite a young man he was apprenticed to the iron-molding trade, and being a hard and conscientious worker he soon mastered all the intricate details of the business. Shortly afterward he became proprietor of a foundry, in which business he was very successful.

He holds the honor and esteem of the community in which he lives, his acts, both public and private, being above reproach. In the administration of the affairs of his office he has shown himself to be eminently qualified for the position, and the kind and courteous treatment which he extends to all in the fulfillment of his duties has made him a universal favorite. He is the youngest man that has ever held that office, but notwithstanding this fact no one can say that the affairs are not perfectly conducted. Mr. Quinn has hosts of friends, among whom he is deservedly popular.

TWO NOTABLE MINES.

The Golden Gate and the Golden Feather, Near Oroville.

The history of the Golden Gate and Golden Feather mines, situated near Oroville, are full of interest. Feather river was the richest gold-bearing channel known in mining history. Where the river could be turned vast treasures were obtained, but owing to lack of capital and engineering skill some of the richest portions of the river bed were left untouched. Major Frank McLaughlin's attention was called to one of these rich strips, now known as the Golden Gate and Golden Feather mines. He prepared maps, plans and requisite drawings of the river and went to England, where he organized two companies, the Golden Gate mine, with a capital of $250,000, and the Golden Feather, with a capital of $1,000,000. Major McLaughlin was made resident manager, and during the past three years an immense amount of work has been done. The portion of the river sought to be worked was in a narrow and almost inaccessible canyon. To reach it roads had to be constructed at immense expense, buildings erected and tools and machinery provided. A rock and crib dam were built, and the poles used in their construction were cut twenty miles away and floated down under great difficulties, and were finally landed on the opposite side from where the flume was to be erected. A high suspension foot bridge had to be built for the workmen to cross. Lack of space will not permit the telling in detail of the difficulties under which the dams, flumes and cribs were constructed, so as to stand the winter floods. The first built were entirely washed away, and large floods impeded progress more than once during their construction. The big flume is over 4000 feet long, and this added to the great canal makes two miles of river bed to work. The bed of the river is covered with a deep deposit of slickens and gravel from twenty to forty feet deep, the result of many years of operations, and this must be removed and the bedrock itself scraped clean, for on this ground has been found the largest quantity of gold in all river mines. The mines are situated in a narrow channel between high and precipitous hills, and there is no place to move the gravel to. So it has been necessary to dig down into the bed and pile the gravel on top of one portion. As soon as this excavation is made and the bedrock on the bottom of the river cleaned of its gold, the great hole thus made is used for a dumping ground for the next section, and it is in this manner that the mines are being worked. Now a permanent dam is being built at the head of the Golden Feather mine which will last for many years and as the gigantic canal instead of a flume will be used, it follows that work upon the mine will continue for a series of years. Only conjectures can be made in regard to the output from these mines, as the amount that is being taken from the river is not made known to the public, nor are the workmen permitted to tell anything about the buried treasures that are now being removed. The general belief, however, based upon well-known facts, in regard to other deposits taken from this river in the past, and from some remarks inadvertantly made by some of the workmen is that large quantities of gold are being taken out. The celebrated Cape claim averaged $500 per linear foot, and at that rate the Golden Gate mine would return its owners $1,650,000. It is believed that the whole section of the two miles of river bed will pay well to work, and the company is building a timber and rock dam that will last for twenty or thirty years. The solid part of this dam will be 140 feet long and added will be a secondary dam 160 feet long. There will be used 500,000 feet of timber and planks, and it will require the excavation of 2500 cubic yards of gravel and about 6000 cubic yards of rock. The cost is estimated at $25,000. The great canal cost about $150,000. Major McLaughlin, the projector of this vast enterprise, of which only the barest outline is given, is well and favorably known in San Francisco to a host of friends to whom he has endeared himself by his geniality, liberality and eminently pleasing social qualities.

AN EXTENSIVE ENTERPRISE.

Operations of the Eureka Lake and Yuba Canal Company.

Probably no better illustration of the disastrous results to the commercial interests of the Pacific coast caused by a discontinuance of hydraulic mining could be had than from a review of the present and past conditions and operations of the Eureka Lake and Yuba Canal Company, consolidated. It must not be supposed, however, that this is an isolated case. On the contrary, the same will apply to every hydraulic mining district in the State of California which has been affected by the injunctions placed upon this character of mining. The casual observer cannot fail to realize that immense amounts of money have been expended in improvements, machinery, etc., all of which is now lying idle, bringing no interest on the capital invested, and the value of which improvements and machinery is steadily decreasing by disuse. This illustration is given so as to show a specific case, and it must be remembered that all other cases are similar. The property of this company comprises 3500 acres of mining land for which it holds United States patents.

Twelve years ago this company was operating six mines, as follows: Consolidated, Columbia Hill, Farrel, at Columbia Hill, Laird, Snow Point, and Boston, Moore Flat. In addition to this, it also operated the American mine at North San Juan.

During the period of the operations of the company its pay-roll averaged about $40,000 a month, having about 400 men directly employed, and giving direct and indirect support to no less than 2000 souls. To give some idea of the enormous operations of this company it may be stated that three large lakes have been constructed in the Sierra Nevada mountains. French lake, the largest of the trio, is situated near Summit City, four miles from Weber lake. The altitude of this lake is 7000 feet and it covers 480 acres of ground. In its construction a dam sixty-four feet high was built. Fancherie lake, near French lake, and Weaver lake are each about 6400 feet altitude. About $400,000 was expended in their construction. Water is carried from forty to seventy-five miles, necessitating the construction of 300 miles of ditches. The original cost of the company's entire plant was $5,000,000, all of which amount may now be classed as idle capital.

It has been estimated that no less than $50,000,000 is now in sight in these mines, and this enormous wealth is waiting for the word when it may be produced and circulated throughout California. The foregoing points regarding this property were gained from Robert McMurray, the representative of the company.

In conversation with him he also gave many facts which are interesting in this connection. During the past eleven years it has been estimated by conservative engineers, well versed in hydraulic mining, that something over $10,000,000 a year have been kept buried by the injunctions on hydraulic mining, thus aggregating as a total nearly $120,000,000.

A careful examination of the damage done to land affected by the debris showed that but $3,000,000 actual damage had been done. This is too broad a showing to need comment. Another fact worth mentioning is that the employes in hydraulic mines received an average compensation of $3 per day, whereas in contrast to this, it is a notorious fact that the laboring element throughout the interior of the State work for wages ranging from $10 to $30 per month and board. Mr. McMurray's labors in Washington in the interests of the present bill before the House have been invaluable and much of the change in sentiment toward the resumption of hydraulic mining must be ascribed to his efforts. Should the bill pass he is certainly deserving of the thanks of the community.

A RICH MINERAL BELT.

The Dutch Flat and Gold Run Districts, Placer County.

The Dutch Flat and Gold Run mining districts are situated on the line

Central Pacific Railroad. A few years ago they were two of the most flourishing hydraulic mining regions in the State. The mines are all idle, having been stopped by the anti-debris injunction. The once populous towns bearing their names are almost deserted, and the few miners remaining eke out an existence by crevicing and cleaning bedrock in the old hydraulic pits.

These districts cover an immense ancient river channel, filled with a deposit of auriferous gravel to a depth of about 800 feet, the deposit between veins being about one-third of a mile wide. In places the top gravel was worked in two benches, one of about 100 feet and one of fifty feet, leaving the bottom or blue gravel remaining. This bottom gravel is known to be very rich, as it has been opened by a long and expensive bedrock tunnel, and worked sufficiently to prove its richness. The amount of gold produced by working the top gravel has been over $40,000,000. This record of the product has been taken from true and reliable sources, and the remaining bottom gravel would produce much more, as the gravel is known to be much richer. The gravel worked is about double the amount remaining; the channel has narrowed down to a width of about 700 feet, the material being much heavier than the gravel that has been washed away. If dams were allowed to be built in the canyons below there could be immense quantities of gold produced.

In referring to the hydraulic district adjoining Dutch Flat the story would be incomplete without reference to D. Munro, the pushing young business man. He is engaged in the general merchandise business, having settled in the community about six years ago. Although competition was close at the time and many large concerns were carrying on business, Mr. Munro, by demonstrating to the satisfaction of all that he had come to do a square business and was there to stay, established a business which has since grown to be one of the most popular and best-known institutions in the county. Mr. Munro is a public-spirited man, every measure deserving the support of the community being earnestly co-operated by him.

Nichols Brothers, the pioneer bankers of the county, have lived in Dutch Flat for nearly twenty-seven years, and during that time have handled most of the gold taken from the hydraulic mines of that vicinity. They have always conducted a careful and conservative business and have the entire confidence of the community. No miner, no matter how ignorant he might have been regarding the value of gold dust, felt a particle of insecurity in trusting to Nichols Brothers for square treatment. The firm have amassed a considerable fortune in the business and are among the most prominent men in the county.

WHITTIER, FULLER & CO.

The Leading Oil and White Lead Concern on the Coast.

What constitutes a solid foundation for the prosperity of a State? Its manufactures. When will California be able to maintain itself against Eastern competition? When she produces the manufactured articles needed by her people. Therefore, her true benefactors are those willing to risk their capital and energies in developing her resources and increasing her manufactures. Pre-eminent in this regard is the great house of Whittier, Fuller & Co., the pioneer manufacturers on this coast in the paint and oil line. Their Pioneer White Lead is the whitest and finest ground, and has greater covering properties than any other made. By its manufacture the output of our mines is used at home, thus saving costly freight charges, and hundreds of miners are kept busy extracting the ores.

What most improves the appearance of a house and preserves it from climatic changes? A good coat of paint. What is

THE BOSTON MINE.

the best paint to use? Whittier, Fuller & Co's Pure Prepared Paint. It is undoubtedly the highest grade of mixed paint made. It is composed of Pioneer white lead, pure oxide of zinc and linseed oil. Containing no adulteration, it is welcomed by first-class painters desiring to do superior work. Being ready mixed, any one can apply it.

What is the safest illuminating oil? Extra Star Kerosene. Why? Because it is made from the finest crude oil yet discovered. It is guaranteed 150 degrees fire test and 49 degrees gravity, combining the highest illuminating power with the greatest safety. Any one trying this oil will use no other. Whittier, Fuller & Co. are sole agents.

This firm manufactures high-grade lubricating oils, greases, etc., especially adapted for mining purposes; also a full line of harness and carriage oils. In the manufacture of their products they use only the finest quality of Pennsylvania premium oils.

They make the well-known brands: Red Star Cylinder Oil, Star Cylinder Oil, Star Dynamo Oil, Star Engine Oil, Red Star Lubricating Grease, Universal Axle Grease and Red Star Harness Oil. These oils are especially adapted to milling and mining machinery, and give universal satisfaction wherever used.

SELBY SMELTING WORKS.

The Leading Establishment of the Kind on the Coast.

The Selby Smelting and Lead Company of California, which was first started in 1865 by Thomas H. Selby, has now grown to such magnitude that it is a well-known institution all over the United States. The works were originally located at North Beach, but being too small to handle the increasing business of the concern a forty-acre tract was purchased at Vallejo Junction, and in 1885 operations were commenced in the new works. Ore is purchased in any quantity and worked at moderate charges. Especially good facilities are offered for the treatment of high-grade, rebellious gold ores which cannot be successfully treated by milling. The institution is complete in every detail and is considered one of the finest in the world.

An adjunct of the works which is noticed by every visitor to San Francisco is the shot tower and lead works on the corner of First and Howard streets. At these works a specialty is made of lead pipe, sheet lead, shot, solder for plumbers and canners, and also for roofing and fine tin work. A feature is also made of lead traps. Another important branch of the business is the manufacture of the standard machine-loaded cartridges, Chamberlin patent. About 100,000 cartridges are turned out per day, in the manufacture of which three kinds of smokeless powder are used, namely, Schultze, E. C., and Wood powder, besides black powder. The head offices of the company are located at 416 Montgomery street, San Francisco. The officers of the company are A. J. Ralston, president, and H. B. Underhill Jr., secretary.

VIRGINIA CITY.

The Center of Nevada's Great Mining Interests.

THE SELBY SMELTING WORKS.

Correspondence of the CHRONICLE.

VIRGINIA (Nev.), May 23.—The view from Virginia City to the eastward is remarkably fine. The eye can sweep in a vista of 180 miles in some directions. To the southeast are the Pinenut mountains about Como. The Twenty-six-mile desert and the Forty-mile desert are also plainly perceptible from C street looking eastward. Far to the right of the Como mountains are the snow-capped summits of the Sierra Nevada. As regards scenery Virginia City has much to boast of, although its immediate environs are desolate in the extreme. It is a city built on a mountain side. In the winter, when this happens to be covered with snow, the view though cheerless is not without interest. The atmosphere is so clear that trees can be distinguished at a distance of thirty miles or more. Virginia City has an elevation of 6205 feet and above the Humboldt plains about 2000 feet. Mount Davidson rises above the city 1622 feet, having a total height above the sea of 7827 feet. Some of the Pinenut mountains in the same range are still higher. In 1859 Virginia City had but two or three houses, and these were stone cabins. A year later the place had quite a metropolitan appearance. Virginia City is to-day the largest city in the State of Nevada and the center of its business and commercial interests.

The residents of the place are very pronounced in their views on the present status of the silver question. W. E. Sharon, superintendent of the Yellow Jacket Mining Company, stated, in conversation with a CHRONICLE reporter, that if an increase in the price of silver was not soon made it would result in the closing down of a majority of the mines in Nevada. "Take, for instance, ourselves," said he; "it is really a toss up whether we keep running or not. For some time past the mine has been kept running with the hope that something would soon be done or some move made whereby a rise in the price of silver would be effected. The cost of production is so great that unless the ore is of remarkably high grade it does not pay to operate it. The people of this State, as well as in the other silver-producing countries, have at last awakened to the fact that they will have to help themselves and the formation of silver clubs has been the result. I think that if all the different clubs of the silver producing districts would combine and work in harmony that much better results would come from their efforts. The people who are directly and indirectly interested in the silver question are now ready to make any move to get relief. As far as bringing the issue into politics and making a straight political issue of it, the result is a question. Of course, if some combination could be effected hereby the votes of the silver States would cut a prominent figure a great deal of good would certainly result. Before this is done, however, the silver States must have some kind of a combination where they will work together."

Mr. Gorham of Gold Hill held very similar views, and stated that it was his opinion that a great field was waiting for a good organizer who could consolidate the interests of the silver States. "We all feel the need of it," said he. "Since the dropping off of the price of silver, thousands of men have been thrown out of employment by the inability of the mine owners to work their mines at a profit. With silver at its old price the effect is too plain to need comment on."

D. B. Lyman, superintendent of the Consolidated Virginia, was of the opinion that the pressure now being brought to bear upon Congress would soon have a good result. The difficulty has been heretofore that but few people realized the enormous extent of the interest which is now suffering. The goldbugs have never yet been able to refute the arguments of the silver men, and the people of the United States are at last beginning to see that we have the right on our side.

Mr. Keating, superintendent of the Best & Belcher, is one of the most convincing talkers in the State on the subject. He has been engaged in mining for a number of years, and has made a close study of the silver question. He is one of the most enthusiastic and hardest workers in the cause, and is one of the leaders among the silver men of the community. He believes that the influence of the silver States will cut a considerable figure in the coming national election, and that if the representatives are firm enough they can secure the desired recognition.

Archibald McDonell, the most prominent mining broker in the State of Nevada, is a hard worker in the cause of free silver. For many years he has lived in Virginia City, being connected with the mining interests first as superintendent and afterward as a mining broker. He is a remarkably well-posted man on the subject and is a strong and convincing talker. He thinks that silver will be brought back to its old standard after the coming election and that the present agitation will be of great value in the fight.

Messrs. Eckley and Wilds, the representatives of the banking institutions of Virginia City, are also hard workers in the cause. They have materially assisted at the formation of the silver clubs and are now looking forward to a combination of all the clubs in the silver-producing countries. Both gentlemen are go-ahead, pushing business men and are very popular in the community.

MINES AND MINING.

There is an impression prevalent outside of California, created in part at least by the people and press of the State, that mining for the precious metals, and particularly gold, is a played-out industry here. This impression has arisen, no doubt, from the phenomenal prominence to which agriculture and horticulture have attained in California, and from the fact that the products of the surface of the soil have attained more recent renown than the treasures which lie buried underground.

It would be idle to attempt here even a review of the history of the discovery of gold in California, for it has become an integral part of the history of the world, taking rank almost with the discovery of America or the Norman conquest of England. It is enough to say that all publicists agree that no event recorded within the limits of authentic history has exerted a more powerful influence upon the economic condition of the civilized world than the finding of a few flakes of yellow metal by James Marshall, a discovery, by the way, which was almost if not wholly an accident. Great wars have changed frontiers, great waves of emigration have altered the social and political status of nations, great epochs of good or bad fortune have exerted a powerful influence upon the destinies of millions, and great men have left their impress upon their native lands for ages and generations, but none of these nor all combined have been so far-reaching or enduring in their results as the discovery of gold in California.

Man, that is, civilized man, might well be defined as a money-using animal. No other animal, no matter how high in the scale of instinct or intelligence, has ever thought out a scheme by which a medium or media of exchanges might be constructed and agreed upon through which the infinite complications of barter might be simplified and reduced to their lowest terms. To measure all things of value by a definite and fixed standard so that every article which every one wants may be set against a quantity or amount of one or two specified things is a feat beyond the intelligence of any animal but man and bespeaks his superiority more conclusively than the discovery of fire, the development of articulate speech or the use of tools. The material prosperity and the substantial progress of a nation are in direct and ascertainable proportion not only to its use of money, but to the stock of money which it has under its control available for proper and necessary uses.

A consideration of these truisms will impress upon the reader the real meaning and effect of the discovery of an admittedly precious metal in quantity in California. The history of 1849 is the history of an impetus in every direction the world over, for gold is like money, it will not remain stationary, but spreads and runs and permeates every country and nation like a universal deluge. No one who is in the slightest degree conversant with the world's history in the nineteenth century can have failed to recognize this important fact.

But 1849 was a good while ago, and the present and future of the mining industry in California is of more interest than the past, glorious and marvelous as it may have been. It is a matter of no difficulty to show, from perfectly authentic sources, that gold mining in California is still an active, wide-awake, lucrative industry, and that if, as the probabilities now are, Congress shall permit the rehabilitation of hydraulic mining, the industry will speedily resume its old-time rank, and a generous rivalry ensue between the golden grain and the golden ore.

The report of the Director of the United States Mint for 1891 estimates the total product of gold in the United States for 1890 at $32,845,000. Of this amount California's output of gold is given at $12,500,000. The next in rank of gold-producing States are Colorado, with $4,150,000; Montana, with $3,800,000; Dakota, with $3,200,000, and Nevada, with $2,800,000. California is thus credited with nearly 39 per cent of all the gold produced in the United States, which is certainly a good showing for an industry not uncommonly believed to have become valueless. This output of gold makes no stir or commotion because it is so largely the result of unpretentious and individual labor. Private owners of mines work away, month after month, with as little parade as though they were farming or woodchopping, but at the end of the year they have added $12,500,000, or over a million dollars a month, to the world's stock of gold.

And this is only one phase of the situation. The report of a commission of United States engineers sent out here to look into the feasibility of reviving hydraulic mining without detriment to the navigable rivers and to the farming and orchard lands of the State asserted that by a conservative estimate the annual yield of gold under hydraulic mining would be $10,000,000, and the report made no attempt to fix a limit to the time during which this output might be expected to continue. The reason for this abstention is very plain. When our visitors go to Dutch Flat next Saturday to be shown what hydraulic mining really is they will be taken into the bed or channel of an ancient river which has been excavated to, say, a depth of one hundred feet or thereabouts below the normal surface of the ground, and they will be told that below the level reached by the hydraulic process a shaft has been sunk vertically for some two hundred and fifty feet, and that all the way down the ground is auriferous, carrying gold in quantity sufficient to make it pay for working by the hydraulic process. With such a showing as that it would be a rash engineer who would venture to put a time limit on the output of gold from the old river channels.

When we consider that the Director of the Mint estimates the world's output of gold for 1890 at only $116,009,000, it requires no argument to show the worth of an annual increment of $10,000,000 for a series of years. We feel certain that if Congress could grasp this statement and appreciate it at its real value there would not be a moment's hesitation about the passage of the hydraulic mining bill, more especially as by the terms of the bill the amount to be expended by the Government in aid of this industry is only about $1,500,000, to be spread over eight years, and this is to be more than repaid by an annual tax of three per cent on the gross output of all the mines which shall avail themselves of the hydraulic appliances supplied by the Government under the direction of the Secretary of War.

The present issue of the CHRONICLE presents the whole situation in a forcible and striking manner. Every effort has been made to give a full and impartial statement of the condition of the mining industry in California and on the Pacific coast, and we commend it to our readers, and particularly to our visitors from the East, as a compendium of information on a subject which cannot fail to interest. We feel that California is entitled to be set right before the world as to the status of her mining industry, and we bespeak the aid of our visitors to assist us in what is certainly a laudable endeavor.

BUTTE CITY.

One of Montana's Great Camps.

Its Remarkable Growth and History.

From Placer Mines to the Largest Bullion Producers in the World.

Correspondence of the CHRONICLE.

BUTTE (Mont.), May 18, 1892.—The history of this place is full of interest, no less to those who made that history than to those who are fond of the romance of those "early days," so dear to the heart of the pioneer.

The first discovery of placer gold on Silver Bow creek was made in October, 1864, between what is now the old town of Silver Bow and Silver Bow Junction. The discoverers, according to Judge Irvine and other old-timers, were Frank Ruff, Bud Barker, Pete McMahon and three others. The locality was on the creek just at its bend, about half way between the two places mentioned. Their manner of working was subject to much inconvenience, owing to the water being too abundant in their excavations and too scarce in their sluice boxes. However, they obviated this difficulty by keeping their works free of water by the use of a China pump, the discharge being into the sluice boxes, into which the dirt was shoveled direct. Notwithstanding all this their diggings paid well. The same fall gold was also discovered at what is now Butte. The discovery was made a short distance below our present Park street and in the vicinity of Arizona street. W. L. Farlin says the discovery was made by a man named Snyder, and that he sold him the lumber with which to construct sluice boxes.

All the first placer mining at Butte was done on the short gulches or ravines at the base of the "hill" and west of it, and the dirt all had to be hauled to water, but it was rich and would stand it. Later on ditches were brought in, but even they were too low on the hillside for many of the claims, and from these the dirt was still hauled either to the creek or to the ditches.

Following these original discoveries at Silver Bow and Butte came discoveries all along the creek between the two places. Then came a time of activity of which it is difficult to give any definite idea. The present mining district was organized with William Allison for president and G. O. Humphreys for recorder, they being the pioneers of the camp. Summit Mountain district was formed later in the fall with W. R. Coggswell as recorder, and when in a few months there got to be too much business for one set of officers to attend to, the district was divided and Independence district created.

Silver Bow town, seven miles below Butte, took the lead by reason of the greater importance of its placer mines at that time, and soon became so much of a town that it was made the county seat. The present county of Silver Bow was then a part of Deer Lodge county. The following year (1865) the county-seat question was settled by a popular vote, and the county officers bundled up their books and removed to Deer Lodge City. One would scarcely suppose from the present appearance of the old town of Silver Bow that it was once a place of over 1000 people, yet such was the case, and even Rocker was a booming town in those days. Butte was the most backward of the three. However, time has recompensed Butte for her halting start by awarding her a wonderful maturity.

What is now spoken of as the old town of Butte was started in the fall of 1864. The first houses (there were two of them) were of logs, and were built one by Ford and Dresser and the other by Humphreys and Allison. They were located at the fork of Town gulch (our present Dublin gulch) at about the point where the road to the Anaconda mine crosses it. A considerable town soon grew up around them. The Ford mentioned was the well-known late Dr. Anson Ford, Butte's first Postmaster and one of her most enterprising citizens for a number of years. The old town, of course, was practically all log houses and not at all pretentious ones either. Many of them are still standing, for when it came to building a permanent city the smoother ground further down and west of the gulch was chosen, and so there was little occasion for tearing down the first log buildings. But Butte stood in that sheltered nook until 1866, when building was commenced upon our present Main street. However, it was not in the nature of a building boom, for Butte even in 1875 was still an unpretentious place.

By the spring of 1867 placer mining was at its height, and gold was being taken out of the gulches and bars all the way from the base of the hills in Butte to below Silver Bow. Butte was now a booming camp, with a population estimated at 2000 or 3000 men, and there were 5000 on the creek. Money was plenty and methods of getting rid of it numerous.

All kinds of business flourished, and active as was the effort for gold, it seemed none the less active to spend it. There was a strong stampede to these diggings, and any one who has ever seen a booming Montana mining camp will readily appreciate the fact that the times were lively, money plenty and its expenditure lavish. As stated above, placer mining extended from Butte to below Silver Bow, all of Silver Bow creek for that distance being taken up with placer claims, each extending 200 feet up and down the creek and reaching from rimrock to rimrock.

The interest in placer mining did not wane until 1869, when, partly on account of a scarcity of water and partly because the richest mines were becoming exhausted and little in the way of new discoveries being made, there began a gradual falling off in the excitement, new-comers were fewer and the out-going tide much greater. But the camp had already produced about $8,000,000 in gold, and if it had stopped right there it would have been safe for a place in history as one of the bonanza localities of the West. But enough had already been done in showing the existence of rich quartz mines here to render it unlikely the district would be suffered to relapse into its original condition after the placer mines were apparently dug out without first thoroughly demonstrating whether or not the rich surface showing of quartz already made would hold out with development. Of course, nobody dreamed of one-hundredth of the truth. Only two lodes were known of at the time, the Original and the Black Chief, with one or two claims outside of these. These were supposed to be the feeders of the placer mines, and there was no thought that the bare hills surrounding Butte in her nestling place were packed with the greatest stores of wealth the world had ever known in so small a compass.

This was not a lively section from 1869 to 1874-75. The quartz mines, for lack of development, did not yet amount to much, although in two or three cases ore of unusual richness had been produced. But very few had any faith in their permanency, for at that time scarcely a quartz mine in Montana had been found to stand development.

The period of quartz development began in 1874, when Butte had dwindled almost to nothingness. But by 1880 the few hundreds who had remained here during the dull days found themselves a part of the liveliest young city of 5000 in the world. The prosperity of the placer mining days had returned with increased activity, and this time it came to stay. The new city

BUTTE CITY IN 1877.

of Butte had drawn away from the site of the old town in Town (Dublin) gulch in 1875 and builded itself on an eminence where a long and active business street stretched up and down the mountain side, while the gentle descent to the east and the west became closely built up with comfortable dwellings, tenanted by a prosperous people engaged either in the chief vocation, mining, or one of the resultant lines of business. It is not necessary to follow the city's growth from that time to this in detail. It is sufficient to say that in the twelve years that have passed the then population of 5000 has been increased to 30,000.

In addition to her population of 30,000 her substantial business blocks, her numerous handsome houses, her immense volume of business, her churches, schools and the other advantages told of above, the city is well supplied with water, and the supply will be quadrupled within a year. It has two electric light companies, a gas company, a perfect sewer system, three lines of street railway—motor, cable and electric—two daily newspapers and a semi-weekly, two telegraph lines, four railroad lines and more coming, a telephone system which does not call forth more than the usual amount of kicking, two messenger service companies, six banks, two fire companies, electric fire alarm system, and, in short, all public conveniences usually to be found in a bustling, wideawake, modern city. Butte has never paid much attention to growing until within the past year, and now that she has commenced in earnest there will be no let up until her population reaches the figure which the importance of her mining interests will justify, and that is something near the 100,000 mark. There are now employed in and around the mines of Butte but little short of 10,000 men, and at the rate new mines are being opened there will be 20,000 men at work here inside of the next two years, justifying the population named above. It is a good place in which to locate in business, and there is no limit to the opportunity for profitable investment in either realty or in mining.

Some of the more prominent mines that have contributed to the growth of Butte are worthy of especial mention.

The great producers of Butte, the ten companies to which almost the total mineral product of the camp is to be credited, are the following: Anaconda Company, Boston and Montana Company, Butte and Boston Company, Parrot Company, Butte Reduction Company, Alice Company, Lexington Company, Moulton Company, Bluebird Company and Colorado and Montana Company.

The product of the first five companies named is copper matte carrying silver and gold, except in the case of the Butte and Boston Company, which also has a bar silver product from its fifty-stamp mill (the Silver Bow.)

The product of the last five companies named, with the exception of the Colorado company, is bar silver. That of the Colorado company is a copper matte, which is of much higher grade in the precious metals than the product of any other copper smelting concern in the district.

The ten companies named operate about forty mines, besides buying and reducing the product of more than that many more, which are worked by their owners or by leasers. Thus almost the entire product of Butte in the form of matte or bullion is sent to market by one of these ten companies.

The Boston and Montana Company derives its ore supply from the same ledges which furnish the great works at Anaconda, though, of course, from different claims on the ledges. The chief producers

Belonging to the Boston and Montana Company are the Mountain View, the two Colusas, the Liquidator, Harris & Lloyd, Moose and Gambetta. During the year several new properties have been acquired by the company. The chief mine of the group is the Mountain View, through which two parallel copper veins of great width and richness run. The shaft on this property has been sunk 100 feet in the past year, giving it a total depth of 1000 feet. The intention is to make it the deepest shaft in the camp, the hoisting equipment being equal to 2000 feet. The East Colusa has not gone any deeper this year, being still 800 feet, but the West Colusa has gained 100, now having a depth of 400 feet. The Harris & Lloyd shaft has been sunk from the 400 to the 500; the Gambetta has doubled its depth, being now 400 feet.

The company has made no change of note during the year in its reduction plant beyond ordinary repairs, and so its capacity remains as it was one year ago — something less than 1000 tons per day. Up to the commencement of 1890 the entire product of ore since the company bought out the Clark smelter has been less than 300,000 tons. The present year the product will be about 200,000 tons, and with the completion of the company's new reduction works at Great Falls, having a capacity of about 2500 tons per day, the product will run up pretty well toward 1,000,000 tons per year. The new works will not be finished, however, until pretty well along in next year. All that is needed to give the Boston and Montana rank alongside of the largest copper producer in the world is sufficient reduction capacity. The mines have already justified the promise made for them by Thomas Couch, manager of the company, as showing capabilities in the production of copper ore unexcelled in the world.

The Butte and Boston Company was organized two years ago and acquired by purchase from the late A. J. Davis a group of thirty-three mining properties, paying for them $1,250,000. The principal mines of this number are the Silver Bow, the Belle of Butte, the two Gray Rocks and the La Plata. One year ago the reduction capacity of the company was about one hundred and fifty tons per day. Of this quantity fifty tons were treated in the company's fifty-stamp chloridizing silver mill, situated at the point of the hill around toward Meaderville, there being an ample output of silver ores from the company's mines for that purpose, the bar silver product of the mill for 1890 aggregating about $750,000. The other 100 tons of the company's daily product was first reduced in the Liquidator concentrator and then smelted in the company's starter for its present plant. Since that time the company has built a new concentrator, with a capacity of 400 tons. Additions have been made also to the actual smelting part of the plant, and from this time on the capacity of the Butte and Boston works will be from 500 to 600 tons per day. These improvements have necessitated additional development in the the mines from which the ore supply is to come. The Gray Rocks have been given 200 feet additional depth, the Silver Bow 100, with corresponding lateral development, while all the other properties have not laid idle by any means. All of these developments have shown large new bodies of a fine grade of silver-copper ores, and the company's capacity for production is now many times its reducing capacity. Mr. Charles H. Palmer, the manager, has had many years' experience as a successful mine manager, but in the Butte and Boston is achieving a success heretofore unequaled even in his experience.

The reduction equipment of the Colorado and Montana Company is about 125 tons per day, though the concentrator is equal to considerably more than that quantity. The company's works are in charge of Henry Williams as general manager, while the mining department is managed by Charles W. Goodale. Two more efficient gentlemen in their respective departments would be hard to find, as is well attested by the great success attending the company's operations. The company's mines are the Gagnon, Original, Butte, Caledonia, Nettie, Hibernial and Burlington. All but about 10 per cent of the ore produced first requires concentration before sending it to the smelter. The principal producer is the Gagnon, which is opened to a depth of 1000 feet. The company treats about 33,000 tons of its own ore per year and also a large quantity of custom ore.

It has been stated in an earlier part of this edition that the matte produced by the Colorado smelter is high grade in the precious metals. The grades of various furnace products, as shown by the newspaper statistics published on the 1st of last January, credits the Colorado Company's matte with carrying about $350 in the precious metals to the ton besides the value of the copper. The precious metal contained in the Butte Reduction Works', matte is about $200 to the ton, that in the Anaconda matte about $130 to the ton, and Boston and Montana matte about $100. So it will be seen that the claims made for the richness of the Colorado product are in a measure substantiated by the figures. The aim apparently is not so much to produce copper as to produce the precious metals, the copper being utilized as a vehicle for gathering the silver and gold. The product of the company's smelter the present year will amount to something over $1,000,000. It is safe to predict, however, from the increased showing for productiveness of the mines owned by the company that there will soon be a very considerable increase in the capacity and yearly output of the reduction plant.

The Moulton Company's chief producer is the Moulton itself, adjoining the great Alice mine on the west, while its

BUTTE CITY IN 1892

tion which
leading edu
and investig
able Facult
systems wil
forms. Soc
of the nobl
and under,
contributed
far prefera
crime, the
owing to t
the govern
without b
intellectual
unnecessar
If you
shall take
Board, or
I re

To th

Henry D
Dear
was pres
to-day, a
was rec
It was
not whic
of the I
mittee, t
an early
It is t
express

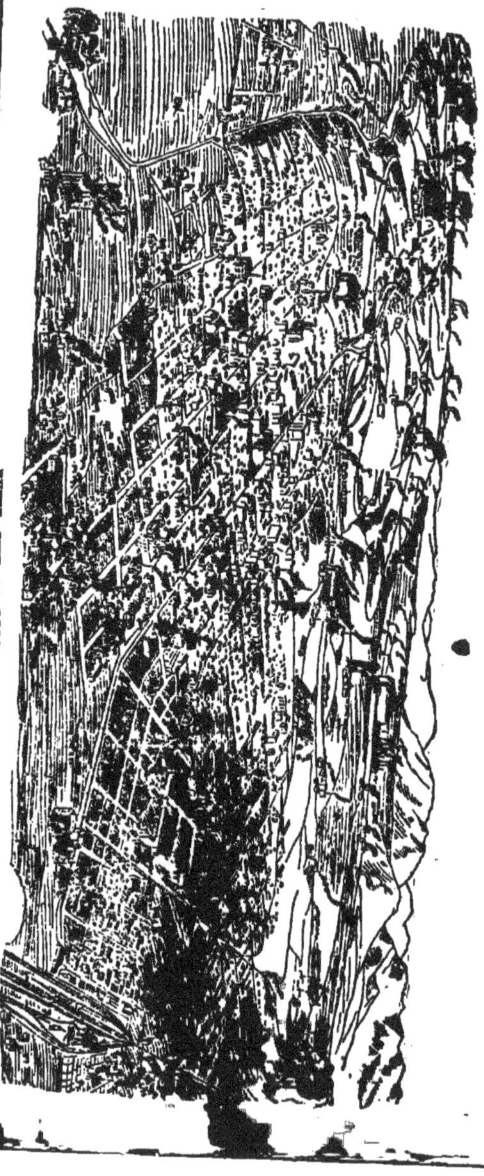

Address

second property in importance is the Poser, which also contains a fine ore-vein. W. A. Clark is president of the company and J. K. Clark its superintendent. The company has an excellent mine equipment and a fine forty-stamp mill. It has not been a dividend payer during the low prices of silver, but if the present agitation results as seems likely in restoring silver to its proper station as money, then the renewal of dividends may be expected from the Moulton. Otherwise, although the Moulton is a great mine, it will be necessary to open bodies of higher grade ore before the stockholders can expect the returns of dividends to any considerable extent. The Moulton's production, as shown by the silver bar shipments, was about $350,000 for the past year.

No account of the finances of Butte would be complete without at least a mention of the strong but ever growing establishment of Hoge, Brownlee & Co. It was founded in 1882 by Messrs. Marcus Daly, R. C. Chambers, William L. Hoge, M. B. Brownlee and F. E. Sargent. These gentlemen saw that there was already room in Butte for another bank, and the success that has come to them amply proves the unerring sagacity which guided them. With such backing failure would have been next to impossible, but their affairs have been so carefully and so ably managed that the realization has far outrun expectation. This bank has ever been found in the forefront when any measure for the general good has been proposed, and to it the city owes much that can never be repaid.

A few words about the Mayor of this city will be read with interest by all, as perhaps no man in Montana is more universally popular. On first meeting Lee Mantle a stranger is at once impressed with the character and individuality which stands forth so prominently in his features. His clean cut face, with heavy brows, and his piercing eyes, is a familiar sight on the streets, and not to know Lee Mantle is to argue yourself unknown. He has, perhaps, the largest personal following of any man in Montana, and his immense popularity has often been demonstrated. Mr. Mantle is one of the class defined by the quotation that "Leaders are born, not made." During the years he has figured prominently before the public in business, politics and general public affairs he has never stood in the background. This must not, however, be ascribed to an excess of assurance. He has never attempted to push himself to the front by such means. Any body of men are quick to recognize the presence of a master mind in their midst, and as a natural consequence his opinion is always eagerly sought. Mr. Mantle is always ready to express his honest convictions on matters pertaining to public interest, and no matter if his views are in direct contradiction to those of his best friends, it makes no difference. Notwithstanding this, however, he is always open to conviction, but his opponent should be well primed and prepared for the contest before crossing swords, for Mr. Mantle is one of the most convincing talkers in the West. He was born in England in 1854, coming to America in 1863 and locating in Salt Lake. In 1870 he went to Idaho, where his first start in business was made. For several years he was agent for the Western Union Telegraph Company and also for Gilmer, Salisbury & Co's stage line. In 1877 he became manager of Wells, Fargo & Co's at Butte, which position he filled to the satisfaction of all concerned. During the early part of 1881 he, in company with a few associates, organized a company and went into the newspaper business. The Daily Inter-Mountain was started, and is now the leading paper in the city. Mr. Mantle was chosen for the position of business manager, and a comparison of the first issue of the paper with those of the present shows that no better choice could have been made. Mr. Mantle was also one of the founders of the Butte racetrack, and up to a short time ago was its principal owner. In politics he has always taken a prominent position, being successively honored by the citizens of Butte. He served two terms as Alderman, twice in the Legislature and was a delegate to the Republican national convention in 1884. His selection as Mayor of the city during the last municipal election is still another incident showing in what respect and trust he is held by the community. Before this election Mr. Mantle threw consternation in the ranks of the mossbacks by coming out strongly and advocating many improvements to the city. Pure water has long been needed, and during the campaign he and his associates formulated a plan whereby this boon could be had. He was met with sturdy opposition, but his personal following and the belief of the people in his judgment proved too strong, and he was elected. He is simply following out his policy, which has placed him in his present position in the community. It is the same with all affairs tending toward the public good. He says that nothing is too good for Butte or the people residing there, and he certainly has shown in the past that he firmly believes this. In fraternal relations Mr. Mantle is also prominent, be being the first grand chancellor, Knights of Pythias, in Montana. He was also president of the Mineral Land convention held at Helena in 1888. Mr. Mantle was especially fortunate in his selection of a business partner, General Warren. Possessing many of the same characteristics as Mr. Mantle, General Warren has made a name and a record for himself which is second to none in Montana. In promulgating and handling large deals General Warren has shown himself to be an adept, his keen foresight enabling him to instantly grasp and master the minutest details of a large enterprise. He is also one of the most popular men in Butte.

In referring to matters pertaining to public interest special mention should be made of H. L. Frank, the wholesale liquor dealer. He is known in Western parlance as a "hustler," and withal one of the most popular men in the city. Whenever any movement is on foot whereby the public interests will be served Mr. Frank is invariably found in the front ranks working hard to push the affair to a successful completion. During his residence in Butte he has built up one of the largest wholesale houses in the Northwest, his business increasing steadily increasing year after year. Mr. Frank has no patience with the specimens so often seen in San Francisco, viz., the silurian. His arguments on this subject are right to the point, and his manner shows that he means exactly what he says. He has determined on Butte as a permanent place of abode, and if the line of policy laid down by Mr. Frank is carried out it will not be long before Butte will be classed as the Chicago of the West.

SHASTA COUNTY.

One of the Old Mining Regions.

Plenty of Mineral Still to Be Found.

Gold and Silver Mines That are Now Being Worked—Good Prospects.

Correspondence of the CHRONICLE.

REDDING, May 20.—Much attention has been drawn of late to the Sunny Hill District, Igo and Ono. The former lies about twenty-two miles in an easterly direction from Redding, the county seat. Mr. Joseph Bell and others have been successfully working their claims, and ores are shipped from there to the smelters, which average from $200 to $250 per ton.

Igo lies to the east of Sunny Hill. The mines here were formerly worked by Alvinza Hayward, but injunctions of the United States District Court have compelled the owners to cease operations. These mines will some day yield large returns to the investor.

From two to four miles west of Igo, in the direction of Ono, lies a belt of mining country that has a large number of silver mines therein. One owned and worked by Mr. Rothwell yields from $500 to $800 per ton in silver. This is the Crystal mine. The Chicago, also a silver-bearing mine, has been worked for the past twenty years with good results. Robinson & Sons also own a number of mines very rich in silver, and within the past six months John Wright has discovered and opened a silver mine, one of the richest and best in the county. Mr. Engle is another owner of some good locations. From all of these mines ores have been shipped to the smelters with good results to the owners. At the present time there are in course of erection works for reducing the silver ores of that district by Yount & Co., these gentlemen being fully of the opinion that this will be one of the best silver mining districts in the State. There is plenty of water power and lumber, and these mines can be worked the whole year round.

About fourteen miles north of Oro we come to the group of mines in the French Gulch district formerly owned by William T. Coleman, and known as the Coleman mine. It is now owned and operated by an English company, William T. St. Auburn being the superintendent. This group of mines has been worked with good results during the past few years.

The Washington mine, the property of E. Lewin, John Souter (manager) and others, has ten stamps continually running. In fact, this mine has been worked since 1852, and has all the time been a paying property. Thomas Green formerly owned a mine in this district. There is a mill on the property, but bad management caused a loss to the owner.

The Gladstone, one of the leading mines of Northern California, is owned by an English company and is under the management of Colonel C. J. Clark. Twenty stamps are continually running with an output of thirty tons per day. They last month crushed 1400 tons. The ore assays high, and the company has made money from the start. Mr. Clark is now negotiating for an electrical plant.

In addition to these mines is the old Highland mine, owned by Mr. Lowdon and others. They run a stamp mill by water power. This mine has produced ore in large quantities that at times has paid very richly. J. W. Conant has a mill on his mining property near by. This mine has been partly opened, and so far as prospected looks very promising and bids fair to be one of the good paying mines of the county.

Frank W. Wheeler has a number of valuable locations on which he has done considerable development work, and near by, just the extreme edge of the

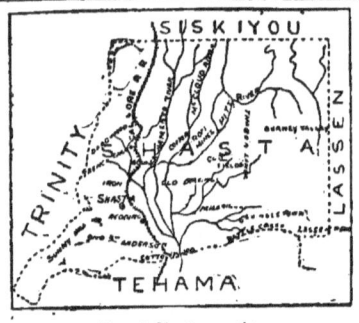

Map of Shasta county.

county, bordering on Trinity, is the celebrated McDonald mine, owned by McDonald Brothers & Franck. This has been and is one of the best-paying mines in the northern part of the State. There are many very promising mines in this neighborhood. Near to and north of Redding is the Scherer Tellurium mine. This is about two miles from Redding and is owned by Peter Scherer and others. Very high grade ore has been taken from it. Close by is a mine owned by Fred Grotefend and others upon which there is a mill. About four miles north of here are the mines and reduction works of A. B. Paul, whose experiments with the new cyanide method, known as the McArthur-Forrest process, are producing good results and have given much encouragement to the miners of Shasta county. These mines are good and paying all the time. A very fine sample of ore was lately brought to the secretary of the Shasta County Miners' Association by Mr. Paul.

A little farther north is a group of mines in the old Diggings district belonging to the Walker Brothers, who have been reducing by the mill process, but contemplate putting in a McArthur-Forrest plant. Adjoining them is the Reid group of mines, with stamp mill, now bonded to a San Francisco company that is about to erect large works. Robert Stevenson, a shrewd Scotchman, is the superintendent, and the mine will be worked for all it is worth. Adjoining is the Central Mine and Milling Company, owned by a New York company, headed by Vice-President Morton. The ledge is large, and ore has been shipped in large quantities to the smelters. Next to this mine is the celebrated Mammoth mine, with a ledge fifteen feet wide, owned by a San Francisco company and under the management of A. J. Morton. Adjoining is the well-known Texas and Georgia mine. A tramway carries the ore from the mines to the mill. R. G. Hart, the owner, is a man of enterprise, and is doing well on his property, which is a very valuable one. Near to him is a group of mines owned and managed by Colonel J. F. Lyons. These are promising well, and much rich ore is on their dump.

Some miles higher up and on the west side of the Sacramento river we come to the Squaw Creek mining district. It would take much space to give a full description of the mines in this rich country, as indeed it would do to justice to any of the districts already mentioned.

The Uncle Sam group of mines is very rich. These mines are owned by the Sierra Butte Mining Company, which has a large stamp mill and tramways running to it from the mines. William James is the superintendent and he is one of the most painstaking men in the country.

The Snyder Mining Company has a large quartz mill. This company has completed tramways from its several mines to the mill, and the ore is delivered at a cost of 10 cents per ton. These mines are under the management of James Barron, one of the best-posted men in the county on quartz mining.

Morton, Bliss & Co. of New York have here a good mine that is unfortunately shut down at present. These mines are very rich, and under a good superintendent would yield largely and be a fine-paying property. In this neighborhood is a large number of smaller mines very promising, but too numerous to mention.

Who has not heard of the wonderful mineral deposits in the Iron Mountain region, about five miles north of the old town of Shasta? By an easy route we arrive at the celebrated Iron Mountain mine, which is under the immediate superintendence of James Sallee. Near by is a mine owned by B. N. Bugbee & Co. of Sacramento and at present under the management of J. M. Gleanes. Adjoining these mines is the Hidden Treasure, which is making a very good showing.

North for the next ten miles, in addition to the gold, silver ore crops out here and there until the Balaklava mine is reached, owned by B. Conroy and others. The ore here is in immense quantities and assays very high in silver.

Of late years but little mining has been done in the Copper City district. A few skilled mining men who understand the business have lately opened up some old claims, and they show promising prospects. The great advance made in the knowledge of quartz mining during the last ten years has given hopes that the previous failures in this part of the county are a thing of the past.

In the eastern portion of the county there has of late been much prospecting done. It has been interfered with, however, by the long and heavy rains. Indications of lead deposits in larger quantities have been found by some experienced miners. How valuable this would be for fluxing purposes need not be told.

Iron deposits of immense extent and of very fine quality, coal, limestone, marble, asbestos and other mineral products of great commercial value go to make up a list that any principality might be proud of.

The Sacramento, Pitt and McCord rivers yield their great strength for water power and immense tracts of land covered by the finest of timber, as yet almost untouched, supply all that the enterprising miner can require.

WILLIAM G. HODSON.

MONTANA'S CAPITAL.

Some Facts About the City of Helena.

A Wealthy and Important Place.

Tributary Railroads Building Up Its Trade—A Bright Outlook.

Correspondence of the CHRONICLE.

HELENA (Mont.), May 20, 1892.—The city of Helena, capital of Montana, lays claim to being the richest city of her size on the continent. She has a greater population than any other point between Minneapolis and Portland, and is the financial, commercial, political and railroad center of an empire of territory. Two railroad lines connect Helena with the outside world, the Northern Pacific and the Union Pacific. The former is her principal outlet, affording her both an Eastern and Western market.

Mining is the chief industry of Montana, and some of the richest mines in the State are situated near Helena. Reduction works have been established at Helena, and branch railways are built to the neighboring camps, thus making the city the point to which may be forwarded the major portion of the State's production. In fact the importance of the place materially depends upon its being the terminus of this network of roads.

Bullion and gold dust are received at the United States assay office in Helena, paying therefor coined gold or currency in return, less the cost of coinage.

The general appearance of the business portion of the city cannot fail to impress the visitor favorably. It is mostly built of brick and stone, giving an air of permanency to the city. The streets in the older portions are somewhat irregular. The miners as they washed their gravel threw it into piles, upon which they built their shanties. These in time were replaced with permanent buildings, and so geometrical lines are somewhat wanting.

Back of the business portion the residences are built with a decided predilection to avoid the plain and climb up the side of Mount Helena. The Courthouse is a striking building and visible from a long distance owing to its being situated on a considerable elevation.

The future of Helena is an assured fact. From the wild placer mining camp of a few years ago it has taken rank as one of the most substantially built inland cities. Its geographical position gives it a commercial importance, and it is one of the best-known mineral centers in the world, while all the valleys around Helena are being rapidly settled by farmers and stockmen. Money is power, and by virtue of her wealth alone Helena can claim her due.

John C. Curtin, the present Mayor of Helena, is one of the pioneers of the State, and withal one of the most popular. He is engaged in the hardware business, which by close application and ability he has built up so that it is now the leading institution of this character in Montana. Mr. Curtin's election as Mayor of the city was nothing more than a deserved recognition of his worth and his standing in the community. Like the majority of merchants in Helena he is interested to some extent in mining, and is doing all in his power to have silver recognized as a precious metal. He has been prominent in the organization of silver clubs in Montana, his work on this subject being especially valuable to the miners. Mr. Curtin says that he is a mossback, at least people call him so, because he chooses to burn coal oil in his store in the nineteenth century. If, however, he should place himself on this standard, the progressive and enterprising men of Montana would be hard to find. Mr. Curtin is confident that the present agitation throughout the silver-producing countries will bring good results, and that the effect will be that silver will stand in the same position that it did ten years ago.

One of Montana's most prominent citizen's is ex-Governor S. T. Hauser, who was born in Falmouth, Ky., in 1834. He studied civil engineering while a young man, and after spending a while in the South started for Montana in 1862 by way of the Missouri river. A large party of adventurers, including young Hauser, went up the river on two steamboats and landed at Fort Benton, then a remote Indian trading post. Their purpose was to cross the Rocky mountains to the Columbia river. They heard at Benton that the Indians had found gold in Western Idaho, but did not have any very definite idea of the geography of the country. Hauser formed one of a party of volunteers to explore the region west of Fort Benton and ascertain if it would be practicable to get wagons through it. The country was found to be open up to the base of the Rockies, and the gold-seekers soon scattered in small parties. Not much was done that summer. There was so little money in the possession of Hauser's companions that by the fall their cook, who was paid $50 a month, had all their cash. It was expected that the party would have to live through the winter on "meat straight," and as Hauser was a good shot he was detailed as hunter. Reports came of the finding of gold at Bannock before the winter was over. Hauser hastened to the new diggings, took up a claim, and worked hard with pick and shovel. His claim was on a sidehill, and he was obliged to haul his dirt on a hide to the water to wash it. The next fall he was a member of the famous Stewart expedition which went to the Yellowstone country prospecting for gold, and fought the Indians in several desperate encounters.

In 1864 Mr. Hauser helped to raise money to pay the expenses of a delegation to Washington to urge upon Congress the division of Idaho and the establishment of a new territory east of the Bitter Root mountains. He went as one of the delegates, in company with W. F. Sanders and Judge Edgerton. They were successful in their mission and the Territory of Montana was established. While in the East Mr. Hauser raised a little money to start a bank with in Virginia City, then the chief town of the Territory. This was the beginning of his career as a banker. In 1865 he organized the St. Louis and Montana Mining Company and erected the first smelter in the Territory on Rattlesnake creek, at the town of Argenta. In 1866 the same company erected the first silver mill in Montana at Phillipsburg.

The discovery of Last Chance gulch had developed the town of Helena, which grew rapidly and became the capital of the Territory. In the same year Mr. Hauser organized the First National Bank of Helena. He thus became the founder of the national banking system in Montana Territory. In 1867 he organized the First National Bank of Missoula, in 1868 the First National Bank of Fort Benton, and in 1870 the First National Bank of Butte. In 1870 he organized the first party which explored the Geyser basin in the present National Park. In 1883 he formed the Helena Mining and Reduction Company, and purchased the works at Wickes and the neighboring mines. He graded the railroad twenty miles to Wickes and turned it over to the Northern Pacific Railroad Company. Mr. Hauser's wife is a daughter of Dr. Farrar and a grandniece of Captain Clarke, the famous explorer of the Lewis and Clarke expedition. He is one of the wealthiest men in Montana and lives in a style commensurate with his vast fortune.

TWIN CITIES.

Grass Valley and Nevada City.

Two Great Quartz Mining Centers.

Prosperous Towns in the Foothills Their Advantages and Attractions.

Grass Valley is celebrated as being the oldest quartz mining town in California, evidences of the work of both quartz and placer mining being seen on every hand. It must not be supposed that Grass Valley stands in the light of a deserted mining camp. On the contrary, the town is now one of the most prosperous of California's interior cities. The population is 7000, and they are about as happy and contented a lot of people as can be found. It was originally settled up as a headquarters for the mines in the district, but the residents of the town have discovered that mining was not the only pleasant and profitable feature of the locality, and the majority of its inhabitants have certainly come to stay. In the first place the location is one of the best in the interior of California, the scenic attractions being of the highest order and the climate the typical climate of California. Being built upon rolling ground the streets are not laid out with regard to the points of the compass but rather to meet the exigencies of the occasion. This feature lends an additional charm to the place. The residents, the majority of whom are comfortably circumstanced, have for years been engaged in beautifying their homes, and it is safe to say that Grass Valley is not behind the procession in this regard. A certain indication of the prosperity of any city or town is the support bestowed upon the local newspapers. Grass Valley has three daily papers, all enterprising and newsy, and the liberal patronage they all enjoy is a credit to the business men and residents of the town.

The mining industry is, of course, the chief industry of the place, many noted mines being now worked adjacent to the town. The fruit industry has also proved to be a profitable one, and on every hand may be found the apple, pear, peach and other fruit trees, bearing heavily and growing thriftily. The original orchards were planted simply for the purpose of supplying the homes, but in late years large shipments are annually made, the industry having grown to large proportions.

As a place of residence Grass Valley possesses many attractions. Besides the beauty of the scenery and the perfect climate, the town has the best educational and religious advantages. The Mount St. Mary's Convent, a school for young ladies, has an enviable reputation throughout the coast, and is considered a leading institution.

The principal mine of Grass Valley is the Idaho, which is also probably the best-known mine in California. Messrs. Edward and John Coleman, who are the principal owners in the property, stated to a CHRONICLE representative that, in their opinion, the mine will be worked for years to come, although a depth of 3100 feet has been reached. The grade of the ore remains about the same, while with the many new improvements in milling and mining machinery the cost of production is materially lessened. Both of these gentlemen have personal supervision of the mine, and much of its success must be ascribed to their careful management. Every precaution against accident is taken, no expense being spared in this regard. The Coleman Brothers are also heavily interested in hydraulic mines near Nevada City. Edward Coleman says that with the resumption of hydraulic mining in California it is his opinion that business of all kinds will be given an impetus, and that no greater boon could be asked for by the merchants and, in fact, all those interested in the financial and commercial interests of the coast. With the enormous amounts of money annually taken from these mines and which he places at from $8,000,000 to $12,000,000, it would create a demand for all classes of merchandise, machinery and would also give lucrative employment to thousands of men. In figuring up the damage caused by the working of the mines he states that the most careful estimates fail to show a damage exceeding $3,000,000. Mr. Coleman is sanguine that the miners will be given the relief asked for, if not in the present Congress in the next one.

Alf Tregidgo, principal owner of the Peabody mine, is also sanguine that the miners will be given the desired relief. Mr. Tregidgo is not a hydraulic miner, but his sympathies are with them, and he is one of their heartiest supporters. During the past two months he has run into a specimen lead in his mine which has yielded enormous returns. The mine was abandoned when he started to work it, and over $90,000 was expended in improvements before one dollar's worth of ore was taken out. The property is now considered one of the most valuable in the State.

NEVADA CITY.

Nevada City is situated four miles north from Grass Valley and is built upon a very similar plan. It is the county seat, and next to Grass Valley is the oldest mining camp in California. Originally the town clustered on the narrow flat on Deer creek, which flows through the center of the place. This space, however, quickly proved too limited and soon the hills in every direction were covered with houses. The streets were not laid out with any particular regard to symmetry, and as a consequence the stranger in Nevada City soon gets bewildered in a seemingly interminable maze. The effect, however, is pleasant. Nevada City supports two daily papers, each being well patronized. The city is incorporated and has two schools of different grades, churches of every description, stores of every kind, four large hotels, an efficient fire department, a good theater, two large foundries and twenty quartz mills. The population of the town is about 6000.

Martin, Muir & Co., who have for some time been engaged in the foundry business, have from a small beginning built up a business which is complete in all departments and recognized as one of the solid institutions of Nevada City. It is the principal foundry in the county, the work turned out giving the utmost satisfaction. The gentlemen are very firm believers in the future of Nevada City, and state that they have come there to stay.

The principal banking business of the county is carried on by the Citizens' Bank, the head office being in Nevada City, with a branch bank in Grass Valley. A conservative business is carried on, and the gentlemen in charge have the utmost confidence and respect of the community. Any matter tending toward the public good is cheerfully supported by this institution, the president, E. Preston, and the cashier, J. T. Morgan, being recognized as two of the most enterprising and progressive men in the community. The bank is the second one established in the county, its predecessor going out of business a short time after the incorporation. Like other citizens in this district, the managers of the bank look forward to a resumption of hydraulic mining with the hope that this session of Congress will afford the necessary relief.

PRECIOUS METALS.

The Part Played by Them in History.

A Steady Supply Means Prosperity.

Commerce Dependent on Their Volume.

Effects of Scarcity in the Middle Ages.

The Demonetization of Silver Sure to Produce Similar Results.

I.

COMMERCE AND MONEY.

The Growth of Trade Depends on the Supply of Precious Metals.

No paper devoted to the mining industry would be complete without a review of the causes that have led to the discrediting of silver.

From the dawn of civilization gold and silver have been esteemed as the precious metals, performing the functions of measuring and exchanging values.

Their volume before the discovery of America was not very great, excepting during the palmy days of ancient Rome. It is estimated by Jacob that during the reign of Augustus, A. D. 14, the Roman Empire contained gold and silver to the value of $1,790,000,000. This was a large amount for the time, and its existence probably explains the great prosperity enjoyed by the Romans for a considerable period after the date mentioned.

The same authority, Jacob, estimates that at the time of the discovery of America the supply of gold and silver in Europe had shrunk to the insignificant amount of $167,000,000.

The effect of this terrible diminution of so necessary an article as money can only be realized from a study of prices. According to Landrin and Roswag a ton of wheat was worth only £1 8s. 6d. during the fifteenth century, scarcely more than one-tenth as much as between 1850 and 1880. An ox could be had for £1; a pound of butter was worth 1 penny; eight pounds of beef could be bought for 6 pence, and so on through the brief list of the productions of the dark ages.

In consequence of the enormous appreciation of the value of money commerce was paralyzed. It is true our histories are filled with allusions to the important trade of the orient, and we are informed by some writers that the discovery of America was in large part due to the attempt of the Turks to monopolize the Levantine commerce, but we may set down much of this talk as mere rhetoric. When an author speaks of the valuable trade in spices during the middle ages we have a right to assume that the glamour of the unknown is over him, for, unless condiments were consumed on an immensely greater scale than at present, the trade in them could not have been large. To-day with our vastly increased capacity for absorbing luxuries and all the enlarged facilities for obtaining them from all quarters of the world, our imports of spices constitute scarcely one thirty-third of the entire volume of our import trade.

The statistics of foreign trade in the middle ages are meager. Indeed one might say there are no statistics of any value for the period anterior to 1492. There is a record that the imports and exports of England in the year 1353 aggregated £414,000, or about 2s. 10d. per capita, but it seems to stand alone, and is perhaps untrustworthy. Our chief data, and that on which historians seem to rely, is composed of fugitive remarks from such writers as the court historian of Louis XI, Philip de Comines, whose pictures of the opulence of his contemporaries have been accepted as gospel.

There was undoubtedly trade with the orient before Columbus sailed from Cadiz to find a new route to the Indies, but it was small, insignificant we may say, by comparison with that of modern times. It could scarcely have been otherwise. The lack of money necessarily operated as a barrier to exchange on a large scale. Men were reduced to the expedient of simple barter, and became shut up, as it were, in their own homes. Population increased slowly, and the ravages of war were repaired with difficulty.

The sluggishness of commerce affected the human mind. Learning was not extinct, but it was confined to a groove. Only the monks studied, and naturally the fruit of their education was not of the kind calculated to nourish human interests. They taught man to lean upon God in all things instead of teaching that God helps those who help themselves. The rearing of a lofty cathedral with many spires pointing heavenward or the creation of an eleemosynary institution was their highest aim, and they succeeded in imbuing the most of the world with their impressions. The few energetic minds that escaped the infection were by no means admired, and the projects of their possessors were viewed askance by those high in authority as well as the multitude.

With the discovery of America came a great change. The stimulus to trade began to exhibit itself early in the sixteenth century, and the impetus given continued until the opening of the present century. In 1583 the total foreign trade of England had increased to £3,980,000, or 15 shillings per capita, nearly six times as much per head as during the reign of Edward III. In 1820 it had increased to £89,000,000, or £5 per capita, the increase per head being sixfold.

Of course with this tremendous increase of external trade there was a corresponding or even greater development of the home trade. Adam Smith, the profoundest economist of his time, about the period of the American Revolution dilates upon the fact that the British home trade was not only larger, but was infinitely more profitable than the foreign.

The expansion of trade after 1492 was by no means confined to England. Indeed, Spain enjoyed the first fruits of the great influx of gold and silver, and her prosperity continued uninterrupted, in spite of the costly wars waged by Charles V, until the revolt of the Netherlands, which was brought about by the religious zeal of Philip II. The Netherlands also felt the stimulus in an extraordinary manner, and for a long time remained the unrivaled traders of the continent.

If we turn to our list of prices we find that an ox which could have been bought for £1 during the fifteenth cost twice as much during the sixteenth and five times as much during the seventeenth century. The price of a ton of wheat had increased eight fold between 1492 and the opening of the eighteenth century.

That these great changes were directly due to the plentiful supply of the precious metals has never been seriously denied by any competent authority. Nor has there ever been a successful attempt to show that the increased price worked an injury to the consumer. On the contrary, the evidence is overabundant that when butter was worth a penny a pound very few pounds of it were made and eaten, and that when wheat was selling at one-twelfth its present price the common run of people were glad to get bread made of a mixture of bran and pounded beans.

But the most convincing testimony is that furnished by the growth of population. That of England had remained almost stationary from the date of the Norman conquest up to the close of the fifteenth century. The three hundred years following witnessed a three-fold in crease. From 1066 to 1528 the addition was only 2,206,000; from 1528 to 1821 it was 7,734,000. The population of France increased slowly from 1328 to 1515, the inhabitants numbering 10,000,000 in the first-named year and 14,000,000 in the last. But between 1515 and 1821 the increase was from 14,000,000 to 30,462,000.

Now, whatever we may think of the Malthusian doctrine, none of us will be disposed to assert that population could increase in this rapid manner without some impelling cause. That cause the unthinking might assume would be found in cheapness, but we have seen that with beef less than a penny a pound, wheat one-tenth its present value, butter a penny a pound, and other things in the same ratio, population in England remained nearly stationary during nearly five centuries. The failure to advance cannot be explained on the theory that the English had not developed mechanical ingenuity, or that they were deficient in the trading instinct, for too many other peoples were, like them, equally lethargic. The few brilliant exceptions, among them the Genoese, the Florentines, the Venetians and some of the cities of Flanders, really emphasize the idea, for their trading voyages were, like those of Jason, in quest of the "golden fleece," and when they brought it home with them and put it into circulation trade received an impetus and prosperity ensued. Indeed careful research on the right lines would probably develop the fact that during the middle ages thrift and prosperity were unknown except where money was reasonably abundant, and we may rationally assume that the countries outside of Florence and Byzantium, where the florin and the bezant circulated and were spoken of with a sort of awe because of their scarcity, had very little money of their own. Indeed that admits of no question, for the meager chronicles of the time show that what little money there was in Europe had been struck in some one of the commercial cities, and that precious few of the coins ever reached the masses.

But when the mines of America were opened the estimated $167,000,000 of gold and silver in Europe soon received enormous additions. Between 1492 and 1829, according to Jacob, the production of the world amounted to $7,308,000,000, while Soetbeer estimates it at $9,360,000,000. Between 1830 and 1888 the latter authority estimates the production at $8,370,000,000, or a total production of $17,730,000,000 from the time of the discovery up to 1888.

How much of this vast production found its way into Europe and what has become of it is a mere matter of speculation. Jacob estimates that between 1492 and 1829 India and China took over a third of the whole output, which would have left Europe about $5,000,000,000, or nearly thirty times as much as she possessed at the close of the fifteenth century. If the absorption of the orient since 1829 has been as great as that estimated by Jacob during the period between 1492 and 1829 Europe would have as share of the whole product over $12,000,000,000. Of course there is no such amount in existence now available for money purposes. An outside estimate would, perhaps, be within $3,000,000,000. The other $9,000,000,000 has gone where the pins go, and there is no more possibility of the sum being recovered for money purposes than there would to restore for consumption the missing pins.

But the object here is merely to show that the effect of the constantly increasing supplies of the precious metals was to enhance prices, and that their enhancement was invariably accompanied by an expansion of commerce and all the indications of prosperity. This can best be done by a condensed table showing the effect of the increased production of gold and silver upon Great Britain, until recently the leading trading nation of the world:

Date.	Precious metals in Europe—Millions.	Population of England —Millions.	The price of wheat per ton.	British foreign commerce—£ millions.	British manufactures—£ millions.
1492	34	4	£1 10s		
1829	1000	24	£12 16s	3	1
1870	1700	31	£11	88	67
				547	174

This table should prove conclusively that the expansion of commerce was due to the increased abundance of money, but there is even more striking evidence at hand. The abolition of the English corn laws was followed by a phenomenal growth of the manufacturing industry of Great Britain, and an equally remarkable expansion of her foreign trade. English free-trade writers, ignoring the fact that other nations during the intervening period have made equally striking advances, have hastily assumed that British prosperity since 1846 was entirely due to the fiscal system advocated by Cobden. But it may only be necessary in this connection to point out that between 1846 and 1888 the world's stock of the precious metals was increased by the enormous amount of $7,750,000,000, or nearly half as much as the total production of the world between 1492 and 1888.

We have seen that during the reign of Edward III, when gold and silver were extremely rare, the foreign trade of Great Britain amounted to only 2 shillings 10 pence per capita. During Elizabeth's reign, seventy or eighty years after the discovery of America, it had increased to 15 shillings per capita. At the opening of the present century it had expanded to £6 8 shillings per capita. From 1800 to 1846 there was a marked change. The superficial free-trade observer has remarked the fact that British trade, considered relatively, declined rapidly between 1800 and 1846, the per capita being in 1830 as low as £3 10 shillings, or little more than half as much as at the beginning of the period. He has tried to explain this by holding the great Napoleonic wars responsible for the stagnation, but such an explanation will scarcely prove satisfactory to students of history, who know that the enormous expansion of commerce from 15 shillings per capita during Elizabeth's reign to £6 8 shillings in 1800 was effected in the face of greater discouragements than the Napoleonic wars. During the seventeenth and eighteenth centuries England was torn by civil dissensions and was embarrassed by foreign wars, and the rest of Europe was in no better posture, but British commerce steadily increased.

Now, there must be some explanation of this extraordinary fact which the historian and political economist fails to furnish, but upon which the tables of the production of the precious metals throw some light. During the seventeenth and eighteenth centuries gold and silver were produced on a tremendous scale. Soetbeer estimates the output during these 200 years at $6,355,000,000. With the opening of the nineteenth century the production began to decline. Between 1801-20 he estimates it to have been $830,000,000, but in the decade 1820-30 it fell to $285,000,000, and between 1830 and 1846 it only reached $620,000,000.

Now, it cannot be a mere coincidence that British trade should have steadily increased during two centuries while the output of the precious metals was increasing, and that it should have steadily declined during fifty years while the production was falling off, only to immediately begin increasing again as soon as the production of the precious metals increased. But let the table tell its own story:

	Average annual production of precious metals.	British foreign trade per inhabitant ....
1601-1700..........	$25,000,000	15 shillings
1701-1740..........	32,000,000	£2 to £2 10s.
1741-1780..........	42,500,000	£3 6s.
1781-1800..........	52,500,000	£6 8s.
1801-1820..........	41,500,000	£3 to £3 10s
1821-1830..........	32,500,000	£3 14s.
1831-1840..........	40,000,000	£4 4s.
1841-1850..........	72,000,000	£6 4s.
1851-1860....,.....	180,000,000	£12 17s.
1861-1870..........	180,000,000	£17 7s.
1871-1890..........	209,500,000	£20 5s.
1881-1888..........	188,500,000	£19 10s.

We leave to the gold monometallist the difficult task of explaining away these extraordinary coincidences and to dispute, if possible, the bimetallic theory that upon the broad foundation of an abundant supply of the precious metals rests the whole superstructure of modern trade. If this foundation is undermined the structure must necessarily become weakened, and if it is too seriously assailed the whole edifice must come down.

II.

THE DARK AGES.

Due to the Failure of the Supply of Precious Metals.

It has been stated that Jacob estimated that the value of the precious metals in Europe at the time of the discovery of the new world was only $167,000,000. This does not mean, however, that the entire amount represented coin. On the contrary, the quantity of money was ridiculously small, and by far the greater part of the $167,000,000 represented the treasures of kings, feudal lords and the churches in the shape of ornaments and vessels. The question naturally arises what became of the vast treasures of the Romans, and the query brings to mind the striking fact, which deserves great prominence in the discussion of the money question, that the Roman empire enjoyed its greatest degree of political and commercial prosperity—for the two are linked together—during the period when Cæsar and the other conque were pouring into Rome the treasures of the conquered provinces. The chief object of Roman conquest was the acquisition of the precious metals, and the great generals of the empire were as proficient in the art of mining as they were in tactics. Every resource at their command was employed to coax gold and silver from the earth, and the value of a conquest was always measured by the amount of treasure it resulted in producing.

It has been mentioned that during the reign of Augustus, A. D. 14, the Roman Empire possessed gold and silver to the value of nearly $1,800,000,000, and this amount was steadily added to during the reign of the wiser Cæsars, who worked the mines of the empire indefatigably. Their besotted and blood-stained successors, no doubt, were as eager to obtain gold as the Antonines, but unfortunately most of their methods were destructive in their results. Their senseless wars and continuous internal discord gradually reduced the supply, until finally toward the fall of the empire gold and silver became exceedingly scarce. It was the observation of this fact that led the historian Alison to dispute Gibbon's theory of the decline of the Roman empire and made him unhesitatingly declare that the paralysis of commerce played a far more potent part than the decay of the military spirit.

It would require too much space to argue these views at length, but the Modern who sees the mighty results of the extension of trade, whose operations are not helped by military movements, will be very apt to dissent from the curious idea that an empire can only grow by warlike methods, and see a more plausible solution of the question in the view of the historian who declares that Roman greatness became extinguished because she failed to provide herself with the necessary medium for expanding her commerce.

Whatever other causes may have contributed to the decadence of the grandeur of Rome it must be admitted that the want of the precious metals was the chief one, for the evidence is overwhelming that when the steady supply of the precious metals ceased trade began to decline and the people of a once civilized world relapsed into barbarism.

This period is known in history as the dark ages, and its chief characteristic is generally assumed to be the obscuration of intellect. But it can hardly be said that there was no intelligence, when the evidence is abundant that mental activity was pronounced, in one direction at least. Of religious discussion there was more than enough. The fires of theology were kept burning constantly and the church had heresies of great magnitude to deal with. It was during the dark ages that the magnificent specimens of architecture which the moderns admire, but do not hope to rival, were reared. Certainly the genius displayed in construction and the art with which these constructions were ornamented do not indicate sluggish minds or absence of energy. If they have any meaning for us it is simply that all man's intellect and energy were made to serve a single purpose, and that was the religious one.

It is part of this discussion to determine what causes brought about such a result, and what broke up the habit of centuries and made men turn their thoughts to things material. We all admit, in a general way, that the effect of the growth of wealth is to turn the thoughts of a people from things spiritual. The Attic philosophers and the historians of ancient Rome saw in the increase of prosperity a menace to the worship of the gods, and formulated the concept that religion cannot long survive the insidious attacks of luxury. May we not accept the converse of this idea and assume that with the decline of wealth after the fall of the Roman empire men's thoughts turned to God and religion became their only occupation. In other words, they became religious because their minds were not distracted by the desire for gain, which is infectious when trade is brisk and brings prosperity in its train.

If this view is correct then the assumption that the stagnation of the middle ages was due to the lack of the precious metals, an abundance of which would inevitably have stimulated trade, cannot be successfully swept aside. No other cause could account for the phenomenon. It will not do to accept the rash conclusions of some writers, who tell us that the chief obstacles to trade in the middle ages were the predatory habits of the feudal lords, for they may be answered by the assertion that the feudal system was the outcome of that isolation which is the most striking feature of non-commercial peoples. Feudalism was possibly because there was no extensive trade. It received its first serious blow from the growth of the free cities, and became extinct when the national idea developed itself, chiefly through the necessity for intercourse which the growth of the free cities created.

Nor can it be successfully maintained that the wars of the middle ages proved a barrier to commerce. The same answer may be made to this contention that was suggested as a reply to the assumption that feudalism operated to restrain trade, namely, that the wars were largely in consequence of the absence of the bonds of trade, which in modern times have been found strong enough to avert many wars.

We must, after thorough investigation, accept the conclusion that trade languished during the middle ages for want of the money metals, and that it only revived after the discovery of America and the consequent influx of gold and silver. The increasing supply soon began to exert its revivifying influence, and before the close of the sixteenth century a revolution of the most far-reaching character had occurred. The writers on the Reformation never trace a connection between the supply of gold and silver and the growth of Protestantism, but the critic will soon discover, if he seeks evidence on the point, that the movement could never have succeeded had there not been a wonderful development of material wealth concurrent with it. Motley, in his graphic description of the struggle in the Netherlands, dimly perceived the fact, and presents many instances of the curious association of religion and trade, and in one place enlarges on the growth of commerce in Holland in the midst of the bitter struggle against Philip II, but he gives the reader the impression that it was the religious enthusiasm of the Dutch that promoted trade, and not, as was really the case, the prosperous condition of trade, due to the causes we have indicated, that made the Dutchmen contend heroically for the right to worship God as they pleased and to govern themselves after their own fashion.

If we are convinced that the decadence of the great Roman empire was almost wholly due to the destruction of its trade consequent upon the scarcity of its precious metals what shall we say of modern statesmen who, ignoring all the teachings of history, deliberately enter upon a financial course which if persevered in must result as disastrously to mankind as did the involuntary closing of the mines in the old world? Our vanity may lead us to imagine that our civilization is too broad to permit the repetition of such a catastrophe, but we should not allow ourselves to be led astray by egotism. The ancients had attained a breadth of culture which only the ignorant underrate, and if the light of their attainments was extinguished ours may be also.

That they are in a fair way to be may be inferred from the fact that the single gold standard idea contemplates the abandonment of the use of silver as money metal. If this criminal effort succeeds more than one-half of the precious metals in existence will lose their money character, and the future supply of the metal available for money purposes will be a constantly decreasing quantity.

If the diminution of the average annual output between 1881 and 1888 of the precious metals caused an almost immediate diminution of British trade, what must be the inevitable result of wholly depriving more than one-half of the supply of the precious metals of their money quality? If the reduction of the output from $209,500,000 per year to $183,500,000 made the foreign trade of England shrink from £20 5s. per capita to £19 10s. per capita what would be the effect of cutting down the annual supply from $188,500,000 to less than $90,000,000? We can only infer from studying the table printed above, which shows that with an annual supply of the metals of $72,000,000 the per capita trade of Great Britain was only £6 4s. to which figure it would soon drop, and perhaps lower.

It must not be lost sight of that the supply of gold is constantly decreasing. According to Soetbeer's tables the average annual production of gold since 1851 has been as follows:

1851-60............................$141,000,000
1861-70............................ 132,000,000
1871-80............................ 120,000,000
1881-88............................ 80,000,000

There is not the remotest probability that this supply will be increased. On the contrary, there is every reason to believe that it will continue to shrink from year to year until it ceases to be worthy the designation, and at no very distant day gold pieces may be as hard to acquire and their purchasing power prove as great as during the middle ages, when a modest piece of the yellow metal would buy a herd of cattle. When that time arrives trade must necessarily be paralyzed and the civilized world may easily relapse into the condition from which it was only awakened when gold and silver from America poured into Europe.

III.

DEMONETIZATION.

The Trickery That Accompanied It and Its Effects.

If the consequences of the demonetization of silver are likely to prove so appalling, why do not men shrink from the danger and strive by every means in their power to avert it? The true answer to this would probably be that people of the world are made up of three classes. The smallest, but most influential, acting on the theory, "After me the deluge"; a second class whose chief characteristic is ignorance, who side with the wreckers of society because they are misled by the organs of the influential classes, and a third whose zeal and numbers are scarcely a match for the cunning of the class who devised the scheme of demonetization and have pushed it to the verge of success.

There are some advocates of gold monometallism who unblushingly declare that they favor a single standard because gold is the only metal having invariability of value and who attempt to excuse their position by asserting that the production of silver had begun to increase on such a scale as to threaten to render the metal valueless. Monometallists of this class are as dishonest as the dollar they advocate, which has not an invariable but has a decidedly and constantly increasing value.

There is another class of monometallists who say that bimetallism is an impossibility and that the ratio of the two metals cannot be determined by law. It is to this latter class that attention will first be directed, as by exhibiting the absurdity of their contention abundant evidence will be incidentally adduced to sufficiently disprove the assumption that gold has a fixed value and to dispose of the barefaced falsehood that the increased production of silver had caused uneasiness and led to demonetization.

Bimetallism is as ancient as civilization. There is no period in the history of man when gold and silver were not equally esteemed as money on some ratio fixed by convention or law. That this ratio has not varied greatly may be inferred from the fact that in the time of Darius (521 B.C.) 13½ parts of silver were regarded as equal to one of gold, and were so declared by law. Twenty-four centuries later the proportion was 15½ to 1. No doubt during the intervening period there were serious fluctuations. What caused them, or whether any serious attempt was ever made by the Romans to preserve an equilibrium is debatable. One thing, however, is certain. There is no record of any civilized people since the Christian era seriously proposing to discontinue the use of silver as money.

The first essays in the direction of monometallism were made by the English, and had for their object the enhancement of the value of the enormous outstanding credits of that nation. J. Thorold Rogers, whose authority is rarely questioned on such subjects, declares that in stock exchange securities at least $10,000,000,000 are known or ticketed as English property. But he adds, "This by no means exhausts the debts owed to people who live or accumulate in the United Kingdom. English capital has gone over the whole world. English houses of business are settled in most countries, and the profits due from them are part of the indebtedness which has to be annually paid." And then he proceeds to remark for the benefit of his English readers "You will see, then, that every year vast amount of cash or property has to imported into England to pay the annual charge of the foreign debt held here."

Perceiving this so clearly is it remarkable that the exponent of the interests the English creditor class should hesitate to give his adhesion to a money policy the inevitable effects of which must be to enormously enhance the value of English credits? Nor is it remarkable that the prospect of the immediate gain should have blinded him and made him incapable of perceiving the future consequences of a fiscal policy closely analogous to that of killing the goose that lays the golden egg.

Assuming the correctness of Roger's calculation that the income derived by the British holders of foreign securities and debts is $500,000,000 annually, we can form some idea of the tremendous temporary advantage Great Britain derives from the discrediting of silver. The effect of diminishing the volume of metallic money being to reduce prices, the English creditors have the purchasing power of their income augmented by an amount equal to that of the decline in the value of commodities. As this decline since demonetization by the United States in 1873 has been estimated to be nearly 40 per cent, the English creditors may be said to be benefiting to the amount of at least $200,000,000 a year by the operation.

But while the creditors are increasing their wealth by the easy process of enhancing the value of their credits by inducing the men who make our laws to limit the legal tender quality to gold, which is constantly growing scarcer, the producer and the debtor—nearly synonymous terms in modern times—are slowly being driven to the wall. The producer suffers directly, because he is aim at invariably a debtor. As the prices of his products decline the work of meeting his obligations becomes more and more onerous, until finally he collapses. While he is being dragged down, if he is an employer, he drags down with him all his workingmen. If any one has any doubt of this he has but to observe the course of trade in England since 1883. The whole period has been one of depression and hardship for the toiler, and the average of profits has not been so low for a century.

It will not do to assert that the workingman shares in the cheapness, for the facts are all against such an assumption. In the first place the nominal wages of labor have not remained stationary but have declined, and continue to decline. But far more important is the fact that owing to the paralysis of trade millions of men throughout the civilized world are out of employment and are unable to obtain the money to buy the necessaries of life. What advantage does cheapness have for the million people in London that English statisticians tell us are hovering on the edge of starvation. If beef is a penny a pound and the miserable toiler is penniless, he must starve unless relieved by charity.

The gospel of cheapness is only preached and honestly believed in by those who are assured of a fixed income. They have reason to approve it, for it means that the dollars they are sure of receiving will purchase more when things are cheap than when they are dear. Political economists of the free-trade school, whose chief stock in trade is abstractions, although they endeavor to demonstrate that cheapness is equally beneficial to the whole of mankind, one and all admit that the surest sign of prosperity is a rising market, and that the index finger which points to depression is low prices or cheapness. John Stuart Mill has noted this, and so has every author entitled to consideration. Men so inconsistent can hardly be looked up to as leaders of opinion.

But even though free traders fail to admit the fact that prosperity and high prices are synonymous, and that cheapness and depression are as securely linked as the criminal with a chain and ball attachment, commercial history affords abundant evidence that such is the case. Why, then, should lawmakers defy reason and adopt a policy which has for its deliberate purpose the cheapening of all commodities? If low prices must bring depression and misery in their train why force on the world a monetary system which, if persevered in, must end in a condition of affairs similar to that which existed in the old world before the discovery of America?

That the object of the gold monometallists is to make money dear and all commodities cheap is indisputable. No scientific explanations or specious assumptions can disguise this object. Demonetization of silver was not resorted to because overproduction of the white metal had disturbed the ratio. That is a falsehood which can be disposed of by a table which shows that when Germany demonetized silver it was at a premium over gold on the established ratio of 15½ to 1 (that of the Latin union), and that when the United States, two years later, at the instigation of Englishmen, imitated Germany, silver was at a premium over gold in this country.

VALUE OF AN OUNCE OF 1000 FINE SILVER.

Calendar year.	Average quotation.	Calendar year.	Average quotation.
1833	$1.297	1861	$1.333
1834	1.313	1862	1.346
1835	1.308	1863	1.345
1836	1.315	1864	1.345
1837	1.305	1865	1.338
1838	1.304	1866	1.339
1839	1.323	1867	1.328
1840	1.323	1868	1.326
1841	1.316	1869	1.325
1842	1.303	1870	1.328
1843	1.297	1871	1.326
1844	1.304	1872	1.322
1845	1.298	1873	1.298
1846	1.30	1874	1.278
1847	1.308	1875	1.246
1848	1.304	1876	1.156
1849	1.309	1877	1.201
1850	1.316	1878	1.152
1851	1.337	1879	1.123
1852	1.326	1880	1.145
1853	1.348	1881	1.138
1854	1.348	1882	1.136
1855	1.344	1883	1.11
1856	1.344	1884	1.113
1857	1.353	1885	1.0645
1858	1.344	1886	0.9946
1859	1.36	1887	0.97823
1860	1.352	1888	0.93987

It will be noted from the above table, which, by the way, was published on the authority of the Director of the Mint, that the value of an ounce of 1000 fine silver was $1.322 in 1872, not three months prior to demonetization by the United States. On the ratio of 15.993 to 1—that fixed by our law—silver was therefore at a premium of over 3 cents an ounce, the value of an ounce of 1000 fine silver at our ratio being $1 29.

How in the face of such evidence can there be any pretense that the object of demonetization was to prevent the imaginary ills that might flow from the overproduction of silver? As a matter of fact, no such pretense was made at the time. Indeed, no pretenses were made at all. The job was done by stealth, those accomplishing it working like sneak-thieves, making believe, while they were consummating an infamy, that they were only effecting some trifling changes in the laws governing the Mint.

It is not a profitable task, however, to discuss this phase of the question. It only shows up in an unenviable light the fact that Americans were overreached by such counselors as the English banker, Ernest Seyd, who furnished the brains for the men who framed the law to strike down silver.

The law demonetizing silver had scarcely been made public when its effects began to be perceived. Silver which was quoted at $1.322 an ounce at the close of 1872 fell to $1.298 an ounce in 1873. From then on it declined until it reached 84 cents an ounce in 1891, the quotation fluctuating about this figure at present.

Nobody heard of the dishonest silver dollar until the dishonest gold monometallists, by demonetizing silver, caused its price to drop. But when in 1876 the agitation for remonetization commenced, the fall of prices sharply calling attention to the job and its consequences, then the gold monometallists hypocritically called attention to the fact that the advocates of bimetallism desired to declare by law that the value of an ounce of silver was $1 29, when the market quotations actually showed that it was only worth $1 15 an ounce.

These dishonest critics ignored the fact that silver, like gold, derives its chief value from its use as money. Had the United States in 1873 deliberately enacted a law declaring silver the only legal tender money, the effect would have been to depreciate the value of gold as compared with silver, and if the whole world at the same time had entered into an agreement to use only silver, an ounce of gold would speedily have purchased less products than an ounce of the white metal.

No sensible man will attempt to dispute the proposition that the value of the two precious metals is determined by their use as money. We make this assertion with the full knowledge that Edward Atkinson has placed himself on record as holding the theory that gold has a fixed value, but such a theory is too ridiculous to receive a moment's respectful consideration. No thing can have value in itself, its value can only be determined by measuring it against something else. Stanley Jevons in his "Theory of Political Economy" has given a concise definition of value, which shows how ludicrous the view of Atkinson. He says:

Value in exchange expresses nothing but a ratio, and the term should not be used in any other sense. To speak simply of the value of an ounce of gold is as absurd as to speak of a ratio of the number seventeen? What is the ratio of the number seventeen? The question admits of no answer, for there must be another number named in order to make a ratio, and the ratio will differ according to the number suggested. What is the value of iron compared with that of gold is an intelligible question. The answer consists in stating the ratio of the quantities exchanged.

Here then is indicated the only mode of ascertaining the value of gold. How much of iron, steel, cotton or any other commodity will it purchase? If it will purchase twice as much to-day of nearly every conceivable article, then obviously its value has increased, and has not remained stationary or fixed. Let us take some figures from the latest abstract of the United States Bureau of Statistics and see whether gold has fixity of value.

In 1872 pig iron was worth $18 a ton; in 1891 it was quoted at $17 52. In 1872 bar iron rolled was sold at $97 63; in 1891 it could be had for $42 56. Steel rails, which were sold for $85 a ton in 1872, are now a drug at $30. Cut nails were $5 46 a keg in 1872; in 1891 their price was $1 86. Bituminous coal which cost $1 84 a ton in 1872 was $2 60 in 1891. Wheat, which was $1 47 a bushel in 1872, tumbled to 83 cents in 1890. Sole leather made a tumble from 25 cents to 16 cents a pound; eggs from 26 to 17 cents a dozen, and so on through the whole long list of productions.

Who will have the temerity in the face of such evidence to repeat the foolish assertion that gold has a fixed value or that a gold dollar is always worth the same? As well might the claim be made that the burglar's jimmy with which he pries open the door of the house he is going to rob is an honest instrument, as to insist that a gold dollar, which will now buy 45 per cent more of almost every commodity than in 1872, is an honest dollar.

The only honest dollar is the one which will purchase as much of an average number of commodities at one time as another. Statistics show that the silver dollar, or rather silver bullion, comes nearer possessing this quality than any other dollar or metal. In support of this statement we submit testimony from the columns of an extreme gold monometallic paper, the New York *Tribune*, and R. G. Dun's Commercial Agency report, which substantially agrees with it. On April 11th the *Tribune* said:

Much current complaint is doubtless due to the fact that prices are low, the range of 200 commodities being about 18 per cent lower than a year ago.

In April, 1891, an ounce of fine silver was quoted at about 15 per cent higher than in April, 1892. We seriously ask which is the dishonest dollar, and whether a gold dollar that will buy 13 per cent more in 1892 than in 1891 is deserving the appellation? If the gold dollar is honest, then black is white.

But the gold standard is not an honest standard, and it would require a violent stretch of the imagination to conceive its chief supporters as really desirous of an honest money. They don't want anything of the kind. What they want is a money which will constantly appreciate in value, so that their credits will be enhancing while they sleep. Their purpose is precisely that indicated by J. Therold Rogers, only they don't put the thing on national grounds. The only nation or religion they know is to increase their riches, and what better means could be taken to accomplish such an object than to have a money whose purchasing power goes on increasing from day to day.

Take the concrete case of the creditor who in 1872 loaned the farmer $5000 with which to improve his property. If the agreed rate of interest was 6 per cent and money values had remained steady, the farmer during the interval would have been able to pay his interest every year by selling 204 bushels of wheat, the proceeds of which at $1 47 a bushel would have realized $300. Finally when it came to lifting the indebtedness he could do so with the proceeds of the sale of 3400 bushels of wheat. But the "honest" gold dollar being the only legal tender, see how the farmer is affected. Wheat drops to 90 cents a bushel, and to raise the $300 to pay his interest account he has to sell 333 bushels instead of 204, and finally, if in spite of adverse circumstances he is able to settle, he must sell 5555 bushels of wheat instead of 3400. Usually he is not able to do so, and the money-lender forecloses and takes his farm.

This is no fancy picture, as the records of the great farming regions of the United States will show. The free traders have raised a hullabaloo about the matter, and charged protection with causing the trouble. But, unfortunately, for their contention the farmers of the United States could pay their debts when prices were high, as they were in 1872, but cannot in 1892, when they are low. Not only has the American farmer felt the pinch of the decline. The trouble has been universal. In England it has caused the value of agricultural land to depreciate fully 25 per cent, and even with the corresponding fall in rentals the British agriculturalist finds it impossible to make a living, and is obliged to send his laborers adrift, and the poor devils emigrate, go to the poorhouse or starve to death.

The trouble is not due to protection and high prices. Low prices are responsible for it ell, and the depression must continue to increase in intensity unless the remedy of resumption of bimetallism is applied.

IV.

MUST BE REMEDIED.

Gold Appreciation Must Stop or There Will Be Trouble.

Can the old order of things be restored with safety? is a question frequently asked, and the inquiry is frequently coupled with another, Would it be fair to resort to bimetallism if the effect of it would be to impair the value of existing credits? Such inquirers have nothing else in mind than the welfare of the owner of money. To them it is immaterial whether commerce flourishes or is depressed; whether men are employed at fairly re-

By Charles H. Oa...

have only waited to see t...
port before bestowing up...
stantial form. Worthy e...
are already at hand, an...
gathering here to-day is t...
not only are an honor i...
givers, but ho...

a most friendly welcome.

www.ingramcontent.com/pod-product-compliance
Lightning Source LLC
Chambersburg PA
CBHW020118170426
43199CB00009B/557